Understanding Statistical Reasoning

Understanding

Statistical Reasoning

*HOW TO EVALUATE RESEARCH LITERATURE
IN THE BEHAVIORAL SCIENCES*

WITHDRAWN

Eleanor Walker Willemsen

University of Santa Clara

W. H. FREEMAN AND COMPANY
San Francisco

A SERIES OF BOOKS IN PSYCHOLOGY

Library of Congress Cataloging in Publication Data

Willemsen, Eleanor Walker, 1938–
  Understanding statistical reasoning.

  Includes bibliographic references.
  1. Statistics.   I. Title.
HA29.W523   001.4'22   73-13647
ISBN 0-7167-0838-8
ISBN 0-7167-0837-X (pbk.)

9 8 7 6 5 4 3 2 1

To my mother

Contents

Preface

This book is intended for use following an elementary statistics course. It has been written for those whose primary need for understanding statistical reasoning arises in their critical reading and interpretation of the research literature of the behavioral sciences. Such critical readers may be engaged in any one of several tasks: they may be advanced undergraduate or graduate students in the behavioral sciences who have been assigned to review the literature in a current research area for a term paper; they may be teachers who wish to prepare lecture notes for a new topic or to write a textbook chapter on that topic; they may be researchers planning a new project, who wish to interpret and integrate the work that has gone before. Understanding written research reports is an essential skill for any serious student of the behavioral sciences. It is hoped that this book will be helpful in developing that skill.

Often in such endeavors, some of the articles being studied are not specifically addressed to the problem that interests the reader. For example, a student might be interested in writing on the development of sex differences in aggressive behavior. A typical source for such a paper might deal with the relation of some antecedent variable like frustration to some measure of aggression. This paper is selected for his reference list by the student, who notes in the abstract that boys were more aggressive than girls in one experimental condition.

A related problem in integrating the available literature in an area is the diversity of different definitions of and measures for terms like "frustration" or "aggression," and many others. Thus, for example, the reader may have at hand several articles that are addressed to the hypothesis that frustration leads to increased aggression, but which suggest conflicting conclusions. Perhaps the frustration-aggression relationship is complex. (It is.) Perhaps the several studies employed different definitions of aggression and different ways of measuring it. Perhaps the studies differ in regard to the populations of subjects sampled, the settings in which observations were made, and the exact circumstances surrounding the frustrating event. If one is interested in the frustration-aggression hypothesis, one must modify one's statement of it to account for all these different circumstances; one will end up saying that it applies only to certain people in certain circumstances. Our student who wishes to study sex differences will also have to take account of this diversity. It may be that the hypothesis applies only to one sex, or that sex differences appear only in certain kinds of circumstances with a particular measure.

In this book the reader will find suggestions for drawing conclusions from the reports of empirical investigations—conclusions that may extend, clarify, or contrast with those of the report's author. These suggestions are not meant to imply that the author's conclusions are usually wrong (though they may be) but instead that they are often not directly pertinent to the reader's term-paper topic or chapter heading. The intent of this book is to help readers become less bound than they often are by the conclusions and interpretations of authors. For this purpose, there are detailed examples throughout the text of statistical tables and graphs, together with discussion of the symbols used in them.

Chapter 1 is concerned with the nature of the statistical conclusion-drawing process. The remaining chapters are devoted to specific techniques and the special problems of interpretation which each entails. Chapter 2 concerns the simpler hypothesis tests for one or two groups; these tests are based on the normal curve or on the t-distribution. Chapter 3 deals with the interpretation of a Pearson product-moment-correlation coefficient and the related regression lines. Chapter 4 is concerned with the analysis of variance and the various tables that summarize the results. Chapter 5 is devoted to multivariate analysis, including factor analysis, multiple regression, and the proper way to inspect a table of Pearson r's. Chapter 6 contains a discussion of interpreting results of nonparametric techniques, which takes up both hypothesis tests and non-Pearson correlation techniques.

The real inspiration to write such a book as this has come from the insistent questions from my colleagues and students. Over the years, there have been very many of these companions in dialogue—

too many to list here. Special mention is due certain colleagues and friends. Lois Stolz, professor Emeritus of Stanford university, Alberta Siegel of Stanford, Leonard Horowitz of Stanford, Hoben Thomas of Pennsylvania State University, Jerry and Nancy Wiggins of the University of Illinois, Marc Nomikos of Berkeley, California. These are all people who have won my admiration for their thoughtful and analytical approach to the ideas they encounter and for their refusal to abandon this approach in deference to anybody's display of statistical virtuosity. I am grateful to them for indirectly convincing me to attempt to write a book to serve the critical reader's need.

Quinn McNemar was my teacher in statistics. He and Leonard Horowitz at Stanford introduced me to the challenge of teaching the elementary course in the subject. I have learned much from, and with, each of them which is reflected in these pages. I am indebted to Leonard Marascuilo for his many helpful comments and criticisms based on several readings of earlier drafts of this manuscript. The responsibility for the manuscript is, of course, mine. Finally, I wish to acknowledge the contribution of Gary Ritchey, who indexed the book.

For their consistently high level of totally nonobjective affection and enthusiasm for any project of mine—even a statistics book—I am grateful to members of my family and to my friends and neighbors in Fresno, California, and in Palo Alto.

Eleanor Willemsen

Palo Alto, California
May 1973

Understanding Statistical Reasoning

The Nature and Description
of Data

The observations made by a modern behavioral scientist usually are presented to his reader as numerical scores. The score assigned to a subject may represent one or more of the behaviors the subject exhibited in the observer's presence, or it may represent a more enduring trait, such as intelligence, self-confidence, or masculinity-femininity. In this chapter, we shall be concerned with (1) how to interpret a given individual score and (2) how to characterize a group of individuals with respect to their scores for a given trait. This enterprise will require that we explore how a behavioral scientist assigns scores to subjects, because the meaning of a score derives from this process. It will also require that we review the methods of descriptive statistics with a view to understanding what information about a group they can and cannot provide.

1.1 The Score

A score is the number that is assigned to an individual subject (person or animal) on the basis of one or more observations made of the subject's behavior.

1.1.1 Measurement Rule

In discussing scores for a group of subjects, an author will often say that a trait, such as dependency, was *measured* for all of his subjects. This measurement produced the set of scores, one per subject, about which the author's subsequent analyses and discussion will center. The implication is that an individual subject's status with respect to the trait under consideration is reflected by his score, X_i. (X is any number obtained by measurement of the trait under study; X_i is a number assigned to a particular individual.)

We can call X_i a measurement if it is a number assigned to individual i on the basis of a rule that can be completely specified. "Completely specified" means that anybody who understands the rule can observe the relevant behavior or attribute of the individual and assign a number to him—and that all observers using the rule would assign the same number given the same observations.[1] Such perfect agreement is rarely attained; in practice, most measurement rules are well (but not completely) specified.

A well-specified measurement rule (1) describes the situation in which observations take place, including any specially important stimuli like test items; (2) defines the behavior to be observed; and (3) describes the procedure to be followed in choosing one, and only one, of the possible values of X for assignment to the observed individual. Thus, given knowledge of a measurement rule and the score, X_i, for some subject, we should be able to infer something about the subject or the subject's behavior or other characteristic.

The number of possible values of X usually is less than the number of subjects who may ever be measured. Thus, it is helpful to think of each value of X as representing a category (figuratively, a "box") into which subjects are classified on the basis of their commonality regarding something they have done, said, or are. Subjects in the same box will be different from each other in many ways. They will even differ with respect to the trait or characteristic being measured. The important thing is that they will be more like each other with respect to the characteristic being measured than any of them is like a subject in a different box. This characteristic of the values of X is contained in the assertion that they are *mutually exclusive*.

In addition to being mutually exclusive, the values of X must have the property that for every subject to be measured there is an available value (box) for him. This property is referred to by saying that the values of X are *exhaustive*. The minimum requirements a set of

---

[1] Such an ideal state of affairs would be referred to by stating that the measurement X had perfect interrater reliability.

X values must meet to say that X is a measurement is for them to be mutually exclusive and exhaustive.[2] This amounts to saying that we can meaningfully classify individual subjects into boxes corresponding to the values of X. "Meaningfully" here refers to the fact that such classification follows a logic (individual subjects in the same box are more alike than those in different boxes) rather than being done at random.

Before a set of scores can be used for the understanding of individual subjects, more than these minimum requirements have to be met. These additional requirements are described in the next section.

1.1.2 Scales of Measurement

A *scale* is simply the set of all possible values for X (the boxes) together with the rule for assigning these values to subjects. There are several types of scales; each type permits different types of statements, about the measured individuals. The most common scales are *nominal, ordinal, interval,* and *ratio;* they represent increasing numbers of requirements the assignment rule must meet.

A *nominal* scale is one that meets only those minimum requirements of Section 1.1.1. Such a scale permits only one conclusion—that the individuals differ from each other with respect to the characteristic measured. Nominal measurement amounts to systematic classification. Classification by hair color is an example. Here numbers, if used at all, could be assigned arbitrarily to the colors.

An *ordinal* scale exists when the assignment rule permits us to assert that the values of X, when arranged in numerical order from low to high, represent increasing amounts of the measured variable characteristic. This is equivalent to saying that individuals occupying adjacent boxes in the numerical order are more like each other than those in nonadjacent boxes. For example, if we rate children for their aggressive behavior, using the numerals 1-5 as possible values of X, we will have an ordinal scale if the children with 4's and children with 5's are more alike than those with 3's and 5's. This is over and above the minimal requirement that 5's be more like each other than any one of them is like a 4, and so forth. In other words, an *ordinal* scale meets the requirements of a nominal scale plus one additional requirement.

If, in addition to sorting individuals (as on a *nominal* scale) and ordering both individuals and categories (as on an *ordinal* scale), the

---

[2]Certain statisticians argue that these minimum requirements do not constitute measurement and that the first characteristic mentioned in Section 1.2.2 must also be present before X can be called a measurement.

measurement rule permits us to make numerical statements about the distance apart of the values of X (the boxes), we have what is called an *interval* scale. Such a scale presupposes that a fixed unit of measurement is being used that has meaning regarding the measured characteristic. For example, children's heights can be measured in inches on an interval scale; height is the characteristic and X is any possible number of inches. An *interval* scale meets all the requirements of an ordinal scale and also has a specified unit of measurement. We can state that individuals who are assigned non-adjacent scores on such a scale differ from each other more than those with adjacent scores and, further, that those whose scores differ by a stated numerical amount on the scale differ by a proportionate quantity of the variable characteristic being measured. This last assertion requires that our numerical unit corresponds to some fixed quantity of the characteristic. For example, a child 56 inches tall is "much taller than" a child 48 inches tall, who is only "very slightly taller than" a child 46 inches tall. By contrast, if we used an ordinal scale, we would only be able to say that the first child was "taller than" the second, who was in turn "taller than" the third child. The statement based on the ordinal scale is clearly less informative about the variation in childrens' height than is the earlier statement based on an interval scale.

Finally, there is the *ratio* scale, rarely encountered in education and psychology. The ratio scale is an interval scale with a known point of origin, or zero point, which permits us to compare scale value X directly. Height, discussed above, is actually such a scale if we agree that the ground on which a child stands while being measured is the zero point. Then we can say that a man of 72 inches is twice the height of a child of 36 inches.

In developmental psychology we often find "age" cited as subject to measurement on an interval scale—in which a year or a month or a day is the unit of measurement; birth is the point of origin; and a score is some number of years, months, or days. Age, however, is not really a psychological characteristic, and it is rare that the author of an article about children wants to restrict his remarks to age as defined above. Usually he wants to discuss some characteristic that changes with age, such as vocabulary size, degree of sensory motor coordination, or cognitive maturity. If *ratio* measurement is claimed (or implied) for these, it must be independently established by showing that the rule for assigning children to values of X for these other (age-related) characteristics meets all the requirements for an interval scale plus the requirement of an arbitrary zero. The idea that zero vocabulary, zero coordination, and zero cognitive maturity are present at birth (age 0) is not self-evident. Trials-to-criterion counts (cited in learning experiments) present the same problem. The onset of the experiment could be the point of origin and a trial could be the unit. But learning does not necessarily start at the onset of an experiment and proceed in trial-sized steps. Ratio scales for psychological characteristics are not common.

If an author does not say which of the above types of scales a given set of X values represents, there are some mathematical techniques that can be used to find out. These techniques are too advanced to be discussed here;[3] however, you can tell something about the nature of an author's measurement procedures by the description given of them. References to systematic classification (often called "coding" in psychological research) implies no more than a nominal scale. References to comparisons by means of rank ordering or comparing individuals in pairs usually implies an *ordinal* scale.[4] If an *interval* scale exists, some definition of a fixed unit would be stated or implied, as in "trials to criterion," which implies that a trial is the unit of measurement. A ratio scale should not be assumed to exist unless a fixed point of origin, or zero point, where there is no amount of the characteristic, can be specified.

For the reader, the important aspect of these distinctions among scales is the relation they bear to the kinds of statements about individuals that we discussed in Section 1.1.1. You should guard against letting yourself (or an author) make descriptive statements about the subjects studied that are not permitted by the logic of the measurement employed. That is, a distinction should be made between (*a*) *statistical* considerations, which are based on properties of groups of scores, and (*b*) measurement considerations, which are based on individual scores. This matter is discussed further in Section 6.4 of Chapter 6.

1.1.3 Reliability and Validity

Statements based on individuals' scores are usually phrased in terms of the variable characteristic—or trait—that the scores are supposed to measure. Thus, we will want to know if the scores measure anything at all in a dependable manner and, if so, whether that "anything" is the trait stated.

The first question is usually answered by referring to the *reliabil-*

---

[3] See Tongerson, W., *Theory and Method Scaling*, New York, Wiley, 1958, Chapter 3.

[4] Additional assumptions popularized by L. L. Thurstone can be made and an *interval* scale presumed to result on the strength of them.

ity of the measurements. Reliability is assessed by determining if two independent assignments of all the individuals in a group to values of X (to their boxes) will result in the same individuals going into the same boxes both times. The independent assignments may be those made (*a*) by two different observers, or (*b*) by one observer or by an objective test at two different times, or (*c*) by two statistically equivalent forms of the same test or by some other assignment procedure.[5] The independence of the two assignments is questionable in methods *b* and *c* above. There will be some carry-over from the first time to the second whether procedures that are identically the same or equivalent are used. This carry-over may result from identical or similar test items and the obvious influence of memory, in which case the danger is overestimation of reliability. If the assignment procedure includes judgments by different persons or by the same person at different times, low reliability estimates often result because of fluctuations in behavior through time or because of different concepts of the behavior—or for both reasons.

In any case, the degree of reliability estimated for measurements is usually presented in the form of a correlation coefficient r (see Chapter 3), which indicates the degree of agreement actually observed between two independent assignments of individuals to scores. Besides noting this value, you should ask yourself whether, on the basis of the nature of the assignment procedure, you should expect the r to be higher or lower than it is.

The issue for you is "are these measures adequate for the purpose to which the investigator wishes to put them or for which I wish to put them?" These purposes may not necessarily require an r at its upper limit, but the conclusion offered should be weighed in light of the reliability reported. For example, one might expect anxiety, as evidenced in a baby's crying, to fluctuate from time to time and situation to situation. Thus a low correspondence (low r) between crying at 10:00 to 10:01 A.M. and at 11:00 to 11:01 A.M. may not be surprising, but it would imply that one should not use 10 to 10:01 crying as an indicant of an infant's enduring tendency to be anxious, or exhibit anxiety behavior. In this example, one would want conclusions based on 10 to 10:01 crying to be quite specific to the circumstances of that day and time. If the investigator wants to call this crying score "anxiety," it is not wrong, but the label "anxiety" adds little to the measure by itself.

[5] The criteria for deciding that two tests or procedures are equivalent are statistically complex and arbitrary at that. A discussion is available in McNemar, Q., *Psychological Statistics*, New York, Wiley, 1962, pp. 148-152.

The second characteristic of a measurement rule we shall discuss is *validity:* the evidence that the rule produces scores that correspond to the trait presumably being measured. For example, if $X =$ the number of verbal requests for help by 10-year-old Johnny in a $\frac{3}{4}$-hour art lesson, is X a measure of the trait "dependency"? the trait "help seeking"? the trait "feelings of inadequacy"? the trait "art ability"? or something other than any of these?

The classical method employed to deal with this question of validity, is to check on the relationship (or lack of it) between X and some *criterion* (an independent measure of the same trait). In this case the investigator chose "the number of verbal requests" as X because he assumed they would be indicators of a 10-year-old child's dependency. Let's image that he therefore obtained (as a criterion) teachers ratings (Y) of dependency from the children's homeroom teacher and sought evidence indicating that the degree of relationship between X and Y was substantial. On obtaining such evidence, he asserted that X was a *valid* measure of dependency as measured by Y. The validity of Y could, of course, be questioned—as it often should be. Notice that the measure X cannot be said to be valid or invalid, but rather that X is *valid for* teachers ratings (Y) and *invalid for*, say, self-ratings of feelings (U).

The logic of this classical technique depends on our having some prior measure, Y, that we are willing to state *is* indeed a measure of the trait in which we are interested. The greatest usefulness of this approach to validity lies in the area of ability and aptitude testing. Here, the criterion measure Y can often be a sample of a well-defined performance skill. If, for example, we are testing aptitude to be a mechanic, the criterion could be $Y =$ "number of mechanical gadgets assembled from written directions in a given time." Or, if we wished to assess sight-reading skills in potential musicians, we could relate our paper-and-pencil test to the number of passages played 100 percent correctly on sight out of a given number of passages.

Returning to the dependency example, we run head on into what is called the *criterion problem*. This is indeed a hindrance to reader understanding, for if he is not sure in the first place what dependency is, how is the reader to assess the information that "help seeking" by children is related to teachers' ratings of the dependency of the children? First, he can formulate (or read) logical explanations. The help seeking of the children and the teachers' ratings both seem to represent some very basic trait that we could call, along with our author, "dependency." Another explanation is that teachers rate children who seek much help as highly dependent. As readers, we may be more interested in this latter idea or in the additional hunch that teachers encourage children they see as dependent to seek help. This last hunch is the same as saying that the author's Y is a *valid* measure *for* the author's X. In conclusion, validity information provided in the classical format must be interpreted in view of what the two measures X and Y really represent. A high relationship between them certainly warrants the statement that either can be called a valid measure of the other. There is no certainty that either of them truly represents the trait stated by the author (or the one postulated by the reader).

This dilemma of interpretation is inherent in validating measures in the classical manner. In recent years, the validity problem has been rephrased and a multiple-measure approach called *construct validity* has been developed. The reader of an article may find himself confronted with this term and associated formal techniques. Even when he does not, however, he will often want to make informal use of the logic to be described here, so that he can better understand a given group of measurements in the context of some purpose of his own.

1.1.4 Construct Validity

Let us return to the dependency example (page 7). In the classical approach, the investigator picked one of many possible criteria and stated or implied that, for his purposes, it was *the* criterion (Y) of dependency. In a construct-validity study, dependency is assumed to be a hypothetical entity (a construct), and all of the variables discussed in the previous section plus, perhaps, several more, are considered to be variables that, in the absence of contrary knowledge, have equal potential for shedding light on the nature of dependency. To keep these indicant variables distinct from one another, subscripts are employed, so that we have, as examples:

X_1 = Homeroom teachers' rating of dependency
X_2 = Number of verbal requests for help in a $\frac{3}{4}$-hour art lesson
X_3 = Test of art ability
X_4 = Self-ratings of "adequacy in my school work"

and in addition:

X_5 = Masculinity-feminity score from interview
X_6 = Male versus female
X_7 = Age at last birthday
X_8 = Attempts to gain approval for art product from peer

We may have some theory about dependency or some prior experience with some of the variables from which we would predict the various relations among them. For example, we may think boys are less dependent than girls and thus expect teachers' ratings (X_1), requests for help (X_2), and attempts to get approval (X_8) to show higher averages for girls than for boys. We may regard feelings about one's adequacy (X_4) as having no systematic relationship with dependency and therefore may expect no significant relation between X_4 and the other variables $(X_1, X_2, $ or $X_8)$.

Perhaps it would be reasonable to say that masculinity-femininity (X_5) is the personality component of sex, in which case we would expect inverse[6] relationships between X_5 and each of X_1, X_2, and X_8, at least for males.

Altogether there are 28 possible two-variable relationships that can be examined for the eight variables, X_1 through X_8. There are additional, more complex relations for more than two variables at a time. Our theory and experience may make predictions about all of them or just some of them. The greater the number of relationships that are predicted, the less likely it is that results consistent with all the predictions could arise by chance alone. This is the crux of *construct validity*.

From theory and experience we predict relations among a number (K) of characteristics. Then we obtain or devise measurements X_1, $X_2 \ldots X_K$ that may or may not truly measure the intended characteristics. To the extent that our predictions are supported, we claim construct validity for both our theory (here, of dependency) *and* for the measurements. If one or more expected relations does not appear, it may be either because the expectation was erroneous or because the measures were inadequate. Further investigation (with the same measures, but different individuals, and vice versa) is required to determine which is the case.

1.2 Distributions of Scores

1.2.1 Data

A single item of information is a *datum;* an example is, "Johnny took eight trials to memorize the list of words." (We usually call such a datum a score; in the example, X equals 8 where X is defined as

---

[6] The low scores on X_5 represent femininity; the high scores represent masculinity. Thus we expect low X_5 scores (feminine scores) to go with high scores on X_1, X_2 and X_8. Conversely, high X_5 scores would accompany low X_1, X_2, and X_8 scores. This matching of high with low, and the converse, is referred to as an "inverse" relationship.

the number of trials required for memorization of a given list.) If we had scores for a group of N children, we would have N items of information; such numerical information is collectively called *data*.

Two kinds of data are usually distinguished. *Discrete data* are those for which only certain specified numbers are permissible scores. (For example, if X = the number of spades in a hand of five cards, X can be either 0, 1, 2, 3, 4, or 5. No other number is a permissible score; values such as 1.12, 3.5, or 7.0 are obviously not possible.) *Continuous data* are those derived from a score whose measurement rule permits all decimal members between, and including, two limits. (An example would be X as the time in seconds required to read a prepared paragraph. All possible decimal numbers between some minimum—for the world's fastest reader—and some maximum—for the world's slowest reader—are possible.)

Scores that provide discrete data are themselves called "discrete scores" and, taken together with the rules for obtaining them, are called *discrete measurement*. Similarly, scores providing continuous data are said to arise from *continuous measurement*. It should be noted that no measuring device can actually be said to allow all decimal values (say, to the fifth decimal place) between two limits. A ruler, for example, is accurate to $\frac{1}{16}$ inch or to 1 millimeter or whatever. In practice, then, data are treated as continuous whenever they are derived from a measurement rule that allows, in principle, all decimal values between two limits. Sometimes data are treated as continuous if the trait presumed to be reflected in them is theoretically continuous even if the measurement provides for only a few of the possible values of X.

1.2.2 The Frequency Distribution

A listing of all of the scores that compose a group of data is not in itself very helpful to one who wishes to understand the pattern of scores in the group as a whole; it is not sufficiently succinct. An especially organized listing, called a frequency distribution, together with certain summary numbers computed from it called *statistics*, provide the necessary aid.

The simplest form of the frequency distributions you will encounter is formed in the following way. The possible values of X are listed in numerical order from largest at the top to smallest at the bottom. Next to each score appears the number of individuals who obtained the score. This latter number is referred to as the F (or frequency) for that value of X. The total of this second column, the F's, should be equal to the size of the group, N, since there are a total of N

Table 1.1 *Frequency distribution of hair colors in a hypothetical group*

Color	Frequency (f)
Blond	125
Light brown	200
Brown	350
Dark brown	325
Black	300
Red	85
Grey	15

individuals who have been distributed among the various X values.

Table 1.1 shows the frequency distribution of hair colors for 1400 hypothetical college students. Since any six distinct numbers could be assigned in any order to the color adjectives, Table 1.1 depicts the frequency distribution for nominal measurements. A bar graph drawn from Table 1.1 appears in Figure 1.1a. A graph for an alternative arrangement of colors appears in Figure 1.1b. Any frequency distribution can be presented in graphic form. Such graphs of discrete or continuous numerical data are often called "histograms"; exactly the same information appears in histograms as in their corresponding tables, but it is in a form that is literally easier to visualize.

Consider next a study in which $X =$ the number of correct answers on a 10-item true-false quiz; this score provides discrete data. A

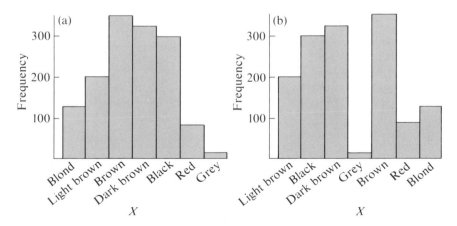

Figure 1.1 *Graphs of the frequency distribution of hair colors.*

Table 1.2 *Frequency distribution of true-false test scores in hypothetical group of 89 pupils*

Score	f
10	2
9	3
8	5
7	14
6	15
5	18
4	17
3	4
2	5
1	5
0	1

frequency distribution of X for a hypothetical class of 89 students appears in Table 1.2. In constructing the graph for a distribution of discrete scores, it is customary to erect a bar for each score over an interval on the horizontal axis that is identified by the score. Figure 1.2a shows the graph for the data of Table 1.2. Figure 1.2b was constructed on the same set of axes as Figure 1.2a, but a line, rather than a bar, was erected over each score. Figure 1.2b is called a "line chart."

When a frequency distribution is made for continuous data, so many values of X are present that intervals of scores are usually used,

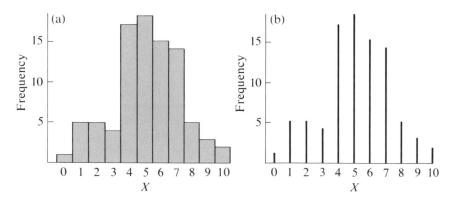

Figure 1.2 *Graphs of the frequency distribution of true-false test scores.*

in place of individual values, in the tabulating process. All intervals are of the same size. In Table 1.3, we see a frequency distribution for $X =$ reaction time in milliseconds (to a small finger shock, let's suppose); 200 people were tested; X was recorded for each individual in whole numbers of milliseconds.

Now, since all decimal values (fractions of milliseconds) are theoretically available as scores, the real limits of an interval are one-half a millisecond below and one-half millisecond above the limits shown. Of course, "one-half" is itself ambiguous—it can be .5000, .49999, or .499999, depending on the degree of accuracy desired in calculations made from the frequency distribution.

The histogram of the frequency distribution of reaction times appears in Figure 1.3. A bar is erected over the midpoint of each interval of scores; the height of the bar is equal to the frequency for that interval.

In both tabular and graphic representations of a frequency distribution for continuous data, it is not possible to discover the exact score of any individual. This detail has been sacrificed in the interests of compact and comprehensible descriptions of the group. The midpoint of the interval serves to characterize the trait status of all individuals in the interval. (For example 124.5 represents all of the individuals in the third interval from the top in Table 1.3, even

Table 1.3 *Frequency distribution of reaction time (X) in milliseconds for 200 people*

X interval	Midpoint	f
140–149	144.5	5
130–139	134.5	5
120–129	124.5	9
110–119	114.5	21
100–109	104.5	40
90–99	94.5	45
80–89	84.5	35
70–79	74.5	20
60–69	64.5	9
50–59	54.5	5
40–49	44.5	5
30–39	34.5	1
20–29	24.5	0
10–19	14.5	0
0–9	4.5	0

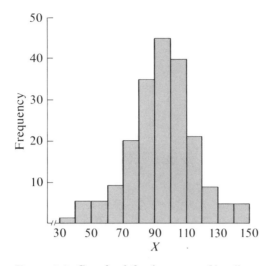

Figure 1.3 *Graph of the frequency distribution of reaction time in milliseconds.*

though one individual may have a score of 120, another 125, and so on.)

The principle here is much the same as that underlying the measurement process, wherein individuals assigned identical scores are not exactly alike, even with respect to the measured trait. Individuals within an interval may be unlike with respect to their scores, but those of adjacent intervals are more nearly alike than those of nonadjacent intervals. The use of the interval *midpoints* serves much the same function in a large group as the use of scores for a very few individuals. That is, a picture is provided by the midpoints and their frequencies of how the scores are distributed.

By looking at the graph, we can discover such information as how bunched together or spread out the scores are, in what score region they are most concentrated, and how fast they taper off on either side of this region. This descriptive information would not be at all apparent from a complete listing of all the scores in the order obtained. Thus we have gained insight—again as the direct result of having sacrificed some specific details.

1.2.3 Interpreting the Shape of a Graphed Frequency

The next logical question to pose is, "now that I have all these data, and I have (or the author has) made a frequency distribution from them, what does it tell me?" Some further computations are

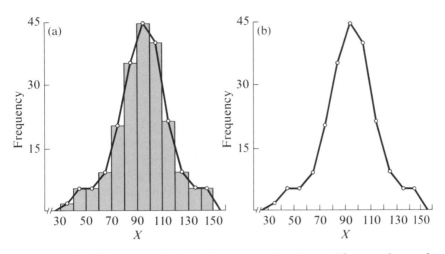

Figure 1.4 (*a*) *Frequency distribution of reaction time with superimposed frequency polygon.* (*b*) *Polygon.*

necessary before all descriptive information can be gleaned; however, the shape of the resulting curve can justify certain statements about the subjects' status as a group on the measured trait. Since the reader is likely to encounter such statements, let us examine their possible bases.

The frequency distributions we shall discuss in this section will be presented in the form of frequency *polygons*, either alone or superimposed on their histograms. Polygons are, of course, geometric figures bounded by any finite number of straight lines. A frequency polygon is bounded by the X axis and a series of lines connecting the midpoints of the bars in a frequency distribution of the bar-graph type. The bar graph of Figure 1.3 is reproduced in Figure 1.4a, with its polygon superimposed on it.[7] Figure 1.4b shows the frequency polygon alone.

Mathematical statistics deals with *probability distributions*, which are idealized versions of these polygons. A probability is a *theoretical relative frequency*—"relative frequency" refers to the frequency (of a given score) divided by N; that is, it is a proportion of the total group. The theoretical relative frequency of a score, that is, its *probability, p*, is a value of the relative frequency of the score ex-

---

[7]Notice that lines are connected from the midpoints *at the baseline* of the intervals immediately to the left and to the right of the frequency distribution to the midpoints of the leftmost and rightmost bars, respectively. This makes the picture a true enclosed polygon.

pected on the basis of theory. In mathematical statistics, this theory is represented by an equation relating p to X. One such equation is, for example, $p = \dfrac{1}{N_s}$, where N_s is the total number of possible X's.

The frequency distribution for a group is considered to be an approximation to the theoretical continuous curve that represents the probability distribution for all possible individuals who might ever be measured. The approximation is accurate to the extent that the group forms a representative sample of all of the individuals who might ever be measured. These latter individuals constitute the *population*, and there would be a very large number of them—perhaps an infinite number—in most cases. The probability distribution can be thought of as the frequency distribution for the entire population. Imagine constructing narrower and narrower bars (with smaller and smaller intervals) for more and more individuals, until, as N approaches infinity, the polygon is visually indistinguisable from a curve. This process is illustrated in Figure 1.5 for a probability distribution that is important to behavioral scientists: it is known as "the normal curve."

All probability curves (whether they are normal curves or not) have an interesting property: the area under a probability curve "adds up to 1.00." To see what this means, it is easiest to refer to a graphed frequency distribution such as that in Figure 1.5a. If the interval

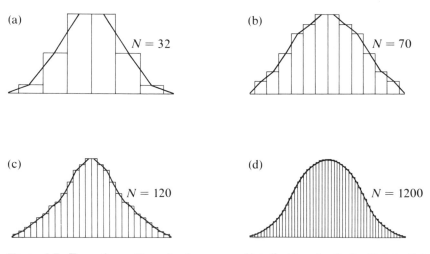

Figure 1.5 *Transformation of a frequency distribution for finite N into the corresponding normal curve for infinite N.*

on the X axis is considered to be *one* horizontal unit, then the area of a given bar will be $1 \times F$ (width on horizontal times height on vertical). The sum of the several areas will thus be the sum of the F's in the frequency distribution. As we know (pages 10–11), this total must be equal to N, so the area under the frequency curve is N. The probability curve is formed by dividing each F by N (relative frequency) and extending this N towards infinite size (theoretical relative frequency). Thus for the area under a probability curve, we have the total of a very large number of f/N's, which is $f_1/N + f_2/N + \cdots$ or $\Sigma f/N = N/N = 1.00$.

This feature of the area under probability curves enables us to find the probability of a value of X being between any given two limiting scores X_1 and X_2. Such a probability would be equivalent to the proportion of total area under the probability curve between X_1 and X_2. These areas are found by methods of the integral calculus; for many of the distributions in common use, such as the normal distribution, they are recorded in standard tables. Tables for the most needed distributions are widely available (in texts, separate books of tables, and elsewhere). You will want to focus part of your attention on the correspondence between an author's statements about his data and the general shape of the probability curve that best describes them.

The normal curve (we see examples in Figure 1.5) is somewhat infamous because many people once mistakenly believed that such a symmetrical frequency curve was the direct result of a natural law that apportions hereditary traits to large groups of persons.[8] Actually, such a distribution often appears for clearly nonhereditary traits. It can be shown mathematically that a normal distribution will appear in a large group of individuals for any set of scores measuring any characteristic that is produced by a large number of subcharacteristics, each of which can be either present or absent (with constant probabilities). Since many genes apparently operate in this manner (each being dominant or recessive), it is not surprising that the hereditary traits they govern are normally distributed in any large group. However, it does not follow that normally distributed traits are hereditary. Many psychological traits that are clearly not hereditary provide nearly normal distributions. A good example would be sociability, for which a test can be based on a person's total number of sociable answers ("agree" or "disagree") to a series of several hundred statements. The statements could be analogous to sub-

[8] See Galton, Francis, *Hereditary Genius: An Inquiry into its Laws and Consequences*, New York, D. Appleton Co., 1970, reprinted; New York, Horizon, 1952.

characteristics, and the sociable response would be either present or absent.

In any instance where the plausibility of a normal distribution is at issue, the answer should be based on the questioner's ability to postulate the existence of, and identify, some of the large number of subcharacteristics. In an instance where a normal distribution seems clearly to be present, it is engaging to try to discover the subcharacteristics. Both of the last-mentioned activities depend on the reader's logical skills and cannot be carried out within the framework of a step-by-step statistical analysis.

Several distribution curves with nonnormal shapes are likely to be encountered frequently in the psychological literature. One type represents a *bimodal* distribution,[9] such as that of Figure 1.6. In this type of distribution, there are two distinct clusterings of scores with tapering off on either side of both clusters. This phenomenon, bimodality, can also be easily detected by glancing down the list of frequencies in a tabular presentation such as that of Table 1.4.

Your interest should, of course, be in the question, "what does this mean?" A bimodal distribution almost always results from the placing together in the same group individuals from two distinct populations. For reaction times of the sort depicted, it may be that infants over four months old react considerably faster than infants under this age. Perhaps there is a rapid growth of neural organization

[9]"Bimodal" is an adjective used to characterize a general shape type. There is *no one single mathematical equation* called "the" bimodal probability distribution.

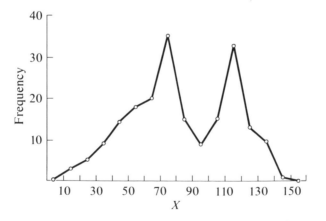

Figure 1.6 *Frequency polygon for a bimodal distribution for reaction times.*

Table 1.4 *Bimodal frequency distribution of reaction time (X) in milliseconds of 200 infants*

X interval	Midpoint	f
140–149	144.5	1
130–139	134.5	10
120–129	124.5	13
110–119	114.5	33
100–109	104.5	15
90–99	94.5	9
80–89	84.5	15
70–79	74.5	35
60–69	64.5	20
50–59	54.5	18
40–49	44.5	14
30–39	34.5	9
20–29	24.5	5
10–19	14.5	3
0–9	4.5	0

during the fourth month. When infants aged two-to-six months old are considered as a single group, the frequency curves showing their reaction times may resemble a juxtaposition of the *two* distinct curves that would be obtained if the two-to-four month olds and the four-to-six month olds were tested separately. In this instance, the variable that operationally defines the two groups is age. Any time bimodality occurs, some such defining variable should be searched for.

It is important to note that in a small group of, say, 10 individuals, apparent bimodality may result from purely random factors, and the group may not really represent two distinct populations at all. The exact size of N at which the two-population idea should be entertained is not dictated by statistical factors alone, but by judgments (including that required for postulating a variable to define the two different populations).

Skewed[10] curves make up the next category of frequently encountered shapes. Any curve may be termed "skewed" if the shape is not symmetrical; that is, if the concentration of scores is located

---

[10]"Skewed" is an adjective used to characterize a general shape type. It does not refer to one (or a family) of mathematical equation(s).

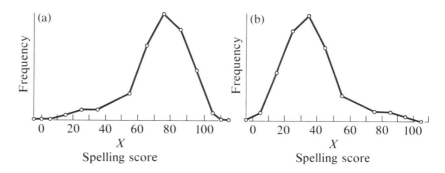

Figure 1.7 (*a*) *Frequency polygon for skewed-left distribution of spelling scores.*
(*b*) *Frequency polygon for skewed-right distribution of spelling scores.*

noticeably to the right (skewed left) or noticeably to the left (skewed right) of the center of the range of scores actually obtained.[11] Notice from the parenthetical terms that the direction of skewing refers to the side of the distribution where the long tail of small frequencies is located. Examples of frequency curves skewed left and skewed right appear in Figures 1.7a and 1.7b respectively. Note that "symmetrical" and "normal" are not synonyms. A symmetrical curve is one that appears the same on both sides of the center of the baseline; A normal curve is a special kind of symmetrical curve. Thus, to say a curve is skewed is to say more than that it is not normal.

There are two principal reasons why a distribution may be skewed: one has to do with the individuals measured, and the other with the measurement process. A distribution of spelling scores that is skewed left could result from a test that is too easy for the population from which the group measured comes, thus producing a majority of scores in the high region. Or it could be that the test is appropriate for the population in the sense that a symmetrical distribution would result for a representative group (sample) but that *this* group is not representative; instead, this group contains a preponderance of individuals with superior spelling ability. A skewed-right distribution is subject to similar explanations: the test may be too difficult for the population, or the group may be inferior in spelling ability to the population.

It should be noted that skewed distributions can result from a combination of measurement and group characteristics. Further, it is entirely plausible that the measurement rule is such that the entire

---

[11]"*Range* of scores actually obtained" refers to the distance along the X axis from the lowest to the highest X-score in the distribution. Algebraically this is equal to the high X *minus* the low X.

population to be measured has a skewed distribution. You should consider whether an investigator provides any clear evidence (even if not identified as being such) that either the measure is not appropriate for the subjects or that they are not typical of the population discussed. Otherwise, you should tentatively assume that the measure has a skewed distribution for the whole population for which it is intended.

1.3 Descriptive Statistics Found From the Frequency Distribution

In addition to inspection of the frequency distribution and its graph or graphs, a reader has recourse to several summary measurements (statistics) that describe the distribution. One can learn to form a mental picture of the distribution graph from these. This exercise is not only possible but desirable, since complete distributions and graphs are often not given in the specialized literature of the professional journals. Indeed, you may be wondering if the frequency distribution exists for anyone other than writers and readers of statistics books! It does. The reasons for not printing frequency distributions are (1) cost and (2) the fact that much of the information they contain can be recovered from the summary statistics. It is this process, the recovery of information from summary statistics, that you will want to learn in this section.

1.3.1 Measures of Central Tendency

Measures of central tendency are the measures that tell where to find the center of a distribution (graphically) or what the most typical score is—there are different ways to define "most typical," of course, and the interpretation of any given measure of central tendency should be made with the associated criterion of "most typical" in mind. Three measures of central tendency will be discussed here. For each of these measures, the criterion of "most typical" is related to a single idea: that a measure of central tendency should provide the *best estimate* available of an individual's score, X, given only the information that X is one of the N scores in a known frequency distribution. The three measures differ in their definitions of "best."

1.3.1.1 The mean. The mean, $\bar{X}$, of a group of scores is simply their sum (symbolized ΣX) divided by the number of them that exist, that is, $\bar{X} = \Sigma X/N$; thus, the mean of 3, 5, 7, and 20 would be $(3 + 5 + 7 + 20) \div 4 = 35 \div 4 = 8.75$. The mean need not be a whole

Table 1.5 *Frequency distribution of X = the number of heads on five coins tossed ten times*

X	f
5	0
4	1
3	3
2	1
1	3
0	2

number nor need it be equal to any one of the values of X permitted by the measurement rule. For example, if I tossed five pennies three times I might get 0, 3, and 2 heads, respectively. The mean of these three numbers is $1\frac{2}{3}$, yet the definition $X =$ "number of heads on five coins" permits only integer scores 0, 1, 2, 3, 4, and 5. Even so, the mean is most descriptive of what happened on the tosses in one sense. Let us see in exactly what way this is true.

Suppose we tossed the five pennies 10 times and obtained the distribution shown in Table 1.5 and Figure 1.8. The mean, $\bar{X}$, from Figure 1.8 is found as $\bar{X} = (0 + 0 + 1 + 1 + 1 + 2 + 3 + 3 + 3 + 4 + 0) \div 10 = 18 \div 10 = 1.8$.[12] Now let us suppose that the ten scores are placed on folded papers in an imaginary hat and that we draw a folded paper out of this hat. We know nothing of the score except that it must be one of those from the distribution given in Figure 1.8. Let us adopt the strategy of guessing that the slip contains the mean, 1.8, on it *even though we know this must be literally false.* If we continue to make this guess for each slip as we remove it, we can determine the average error we make. In each case the error will be equal to the actual score written on the slip minus 1.8, the mean. Sometimes we will overestimate, whereas other times we will underestimate; therefore, some errors will be positive and some will be negative. There will be 10 such errors that we can sum and divide by 10 to obtain average error. This error would be $1/10\,[(0 - 1.8) + (0 - 1.8) + (1 - 1.8) + (1 - 1.8) + (1 - 1.8) + (2 - 1.8) + (3 - 1.8) + (3 - 1.8) + (4 - 1.8)]$. Since the bracketed sum contains the mean subtracted $N = 10$ times, plus a series of scores that total to

---

[12] For larger N's the calculation is facilitated by writing each X score just once and multiplying it by its frequency. Thus in the example

$$\bar{X} = \tfrac{1}{10}[0 \times 2 + 1 \times 3 + 2 \times 1 + 3 \times 3 + 4 \times 1 + 5 \times 0] = 1.8$$

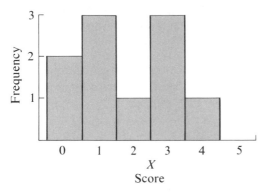

Figure 1.8 *Frequency distribution of* $X =$ *the number of heads on five coins for* $N =$ *10 tosses.*

18 (or ΣX, in general), we arrive at the sum of the individual errors as $[18 - 10 \times 1.8] = 18 - 18 = 0$. In general the sum of the errors incurred by guessing $\bar{X}$ for every score in a distribution is $\Sigma X - N\bar{X} = \Sigma X - \Sigma X = 0$. It follows at once that the average error incurred by using $\bar{X}$ as "best guess" is $0/N = 0$.

The point of all this is that, given no information about an individual score except that it came from a known frequency distribution, the best guess as to what it could be is the mean. This is true provided that having the average error equal to 0 is the criterion of "best." The mean is the measure of central tendency that makes the average difference between a score and itself equal to zero. Any given difference may be large; when a score is present that is much larger or smaller than any of the other scores, the mean is often made much smaller or larger than it would otherwise have been. This fact is referred to in textbooks by statements to the effect that the mean is "sensitive to extreme scores."

1.3.1.2 The median. Another measure of central tendency, often used with ordinal data, is the *median*. The median is defined as the point at or below which fifty percent of the scores fall. When the scores are arranged in numerical order, this would be the middle score for an odd N and the number half way between the score $N/2$ and the score $N/2 + 1$ for an even N. When N is large, the median may be computed from the frequency distribution; this technique is described in basic texts.[13]

In the last example (Figure 1.8 and Table 1.5), the median is 1.5.

[13]See, for example, Hayes, W., *Statistics*, New York, Holt, Rinehart, and Winston, 1968.

This was found in the following way: The scores were arranged in increasing order as 0, 0, 1, 1, 1, 2, 3, 3, 3, and 4. Individual $N/2$ is the fifth individual, and his score is 1. The next score $(N/2 + 1)$ is 2, and the median is therefore half way between these two, or 1.5.

Notice that the median is smaller than the mean, which was found to be 1.8. This is characteristic of any distribution that is skewed right, as is our example. Conversely, any distribution skewed left will characteristically have a median larger than the mean. An exactly symmetrical distribution will have the mean equal to the median. Thus the relation of the mean to the median tells us whether or not a distribution is asymmetrical, and, if so, the direction of its skew.

This property of the mean-median relation highlights the fact that the mean is affected more than the median by a single instance of a score that is extremely large or extremely small relative to the others.

The median is most typical of all the X scores in the sense that it minimizes average absolute distance, $|X\text{-Mdn}|$, between itself and the scores. It is the best bet for X given no knowledge except that X came from a stated distribution, provided that one accepts minimum average $|X\text{-Mdn}|$ as one's criterion of "best." Note that the mean includes the numerical value of every X score in its formula, whereas the median is determined only by the middle one or two scores and N. Any score on one side of the median that can be changed without putting it on the other side of the median may undergo this alteration without changing the value of the median. Thus, for any given distribution, the choice between median and mean for "typical value" depends on one's judgment about whether the typical value should or should not reflect slight numerical fluctuations.[14]

1.3.1.3 The mode.   Sometimes the most frequently obtained score (for an interval distribution, the midpoint of the interval with the greatest frequency), is cited as the "most typical" score. This measure of central tendency, *the mode*, is the best guess for an individual score according to the criterion that its use maximizes one's chance of guessing correctly in a given instance. For a reasonably large N, this chance is always small, but, since it is f/N for each value of X, it

---

[14] There is some feeling among statistically oriented psychologists that the median should be used with ordinal data since slight numerical score differences are not likely to reflect significant trait difference. The mean, they say, should be reserved for measurements on interval and ratio scales, since numerical fluctuations on such scales are supposed to indicate genuine differences in trait status. A discussion of these considerations may be found in Siegel, S., *Non-parametric Statistics for the Behavioral Sciences*, New York, McGraw-Hill, 1956, Chapter 2.

is indeed greatest when *f* is greatest. The mode is literally *the* most typical score.

1.3.1.4 Summary: Central tendency.   These three indicators of central tendency or typical value each tells us something different about a distribution. When one encounters a mean or median or mode, these variations in type of information should be kept in mind. However, because of the ease of its algebraic manipulation and the other (advanced) mathematical properties it has, the mean is the most often used measure of central tendency. For most descriptive purposes, its use is adequate. When the median is also given, the reader can infer whether or not the distribution is skewed. The mode is the most quickly noted indicator of central tendency; beyond that, it is of little use.

1.3.2 Measures of Variation

1.3.2.1 The purpose of measures of variation.   In addition to a measure of central tendency, we need a statistic that amounts to a brief statement about how far away from each other and from the center of the distribution the bulk of the scores fall—the phrase "far away" can be thought of as distance on the baseline of a frequency distribution graph. This second statistic, together with the statistic for central tendency, should enable the reader to visualize the location and the spread—narrow or wide—of the graphed frequency distribution on the *X*-score axis. Abstractly, the statistic of variation should indicate to the reader just how typical the typical value (measure of central tendency) really is. Large variation suggests that the measure of central tendency is not so very typical for many of the scores, whereas small variation suggests that it more nearly is typical. The simplest measure of variation to determine is the *range*, which is the difference between the highest and lowest scores in a distribution. If this value is large, the distribution graph is wide, encompassing nearly the entire series of possible values of *X*, as in Figure 1.9a. Intermediate and small values of the range correspond to intermediate-width distributions and narrow distributions, respectively. An example of the former appears in Figure 1.9b and of the latter in Figure 1.9c, respectively. The most commonly encountered measure of variation, the *variance* (and its square root, the *standard deviation*) uses more of the scores in its calculation than does the range; that is, it reflects more of the information in a distribution than does the range. It is for this reason, and because of certain of its mathematical properties, that the variance is the most commonly encountered statistic for variation.

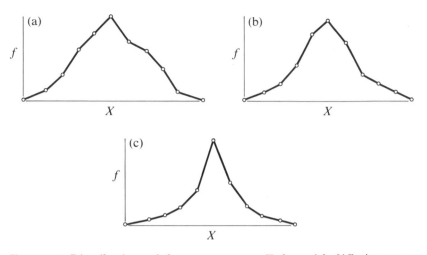

Figure 1.9 *Distributions of the same measure X, but with differing ranges: (a) "wide"; (b) "middle"; (c) "narrow".*

1.3.2.2 The variance. The variance (S^2) of a frequency distribution is defined by the expression

$$S^2 = \frac{\sum (X - \bar{X})^2}{N} \tag{1-1}$$

The capital sigma, Σ, stands for "sum of" and is meant in this instance to refer to a sum of N terms,

$$\sum (X - \bar{X})^2 = SS = (X_1 - \bar{X})^2 + (X_2 - \bar{X})^2$$
$$+ \cdots + (X_i - \bar{X})^2 + \cdots + (X_N - \bar{X})^2 \tag{1-2}$$

This sum is often called "the sum of squares," abbreviated SS. Each parenthetical term in SS represents the squared difference between an individual's score and the mean, $\bar{X}$. Thus it represents the square of the *error* incurred by guessing (estimating) the mean, $\bar{X}$, for that individual. Squaring removes the plus and minus signs; thus, each of the squared errors and, hence, SSs are always positive.

Mathematical statisticians have studied the formula for S^2, and they have shown that estimates provided by Equation 1-1 are too small. This *bias*, as it is called by mathematical statisticians, can be minimized by using $N - 1$ in place of N in the denominator of Equation 1-1. Later, we will have more to say about $N - 1$ and why it is said to stand for the "degrees of freedom" associated with S^2.

When *SS* is divided by $N - 1$, we have, as in the variance, a kind of average or mean of the squared errors. This is why a calculated variance may be encountered under the name "mean square error"—the "d" was deleted from "squared" somewhere in the history of the term's usage.

As an example, let us calculate the variance of the distribution in Table 1.5. First, recall that the mean, $\bar{X}$, was 1.8. Then $SS = (0 - 1.8)^2 + (0 - 1.8)^2 + (1 - 1.8)^2 + (1 - 1.8)^2 + (1 - 1.8)^2 + (2 - 1.8)^2 + (3 - 1.8)^2 + (3 - 1.8)^2 + (4 - 1.8)^2 = (-1.8)^2 + (-1.8)^2 + (-0.8)^2 + (-0.8)^2 + (-0.8)^2 + (0.2)^2 + (1.2)^2 + (1.2)^2 + (1.2)^2 + (2.2)^2 = 3.24 + 3.24 + .64 + .64 + .64 + .04 + 1.44 + 1.44 + 1.44 = 12.76$, and the variance, $S^2 = 12.76/10 = 1.276$.

1.3.2.3 The standard deviation. Because of the squaring done to eliminate minus signs (and for other mathematical reasons), the variance is expressed in units that are the square of the units of measure X reflects. In the example, this would be numbers of heads squared. (More intuitively, we could note that the variance in peoples' heights would be expressed in square inches.) In order to make use of this measure of variation with reference to distance on the X axis in a frequency distribution graph, the square root of the variance is employed. This statistic is called the *standard deviation*, symbolized S. Of course $S = \sqrt{S^2}$ where S^2 is defined in Equation 1-1.

1.4 Standard Scores

Given a set of observed scores for which the mean, $\bar{X}$, and standard deviation, S, are known, the X's may be converted algebraically into a distribution of scores that uses the standard deviation, S, as the unit. These scores are called *standard scores* and are symbolized by the lower-case letter z. The definition formula is given as

$$z = (X - \bar{X})/S \tag{1-3}$$

This change from X's to z's is what is known as a "linear transformation" because the equation relating z to X is that of a straight line; that is $z = BX + A$, where B is $(1/S)$ and A is $(-\bar{X}/S)$. Also $X = (z) S + \bar{X}$ which shows how S functions as a unit with z indicating *how many* S's above or below the mean X is. If z is negative, then X is below the mean, whereas, if z is positive, X is above the mean.

There are two principal uses of the conversion to z-scores, both of which (of mathematical necessity) leave the shape of the distribution exactly as it is for the X's.

One use of z-scores relies on the fact that, whatever its form, a distribution of z-scores has mean $\bar{X} = 0$ and standard deviation $S = 1$.[15] Since this is so, the area between any two values of X may be determined from a table of areas for any probability distribution purported to characterize the X's. Without z-scores, we would need an enormous number of tables for all the X-score means and standard deviations.[16] With z-scores, we can get along with one table for each different distribution shape. The table of the normal curve for $\bar{X} = 0$ and $S = 1$ appears in Appendix A. Columns are provided showing the areas (probabilities—see Section 1.2.3) between the mean and tabled z-scores and also the areas beyond the tabled z's. To use the table, investigators simply convert their X's into z's and proceed.

[15] This is shown by algebra, as follows:

$$\bar{z} = \frac{\sum \left[\frac{X - \bar{X}}{S} \right]}{N}$$

Substituting the formula for z in place of X in the formula for the mean

$$= \frac{\frac{1}{S} \sum \left[X - \bar{X} \right]}{N}$$

Removing the constant, S, from under the summation sign

$$= \frac{1}{SN} \sum X - \frac{\bar{X}}{SN}$$

Separating the two terms

$$= \frac{N\bar{X}}{SN} - \frac{N\bar{X}_X}{SN}$$

Substituting $N\bar{X}$ for $\Sigma \bar{X}$, and for ΣX which $= N\bar{X}$ by definition

$$= \frac{M_X}{S} - \frac{M_X}{S}$$

Canceling the N's

$$= 0$$

Subtracting

and

$$S_z^2 = \frac{\sum \left[\frac{X - \bar{X}}{S_X} \right]^2}{N}$$

Substituting $z = (X - \bar{X}_X)/S$ and $\bar{X}_z = 0$ into Formula 1-1

$$S_z^2 = \frac{\frac{1}{S_X^2} \sum [X - \bar{X}]^2}{N}$$

Removing the constant from under the summation sign

$$S_z^2 = \frac{1}{S_X^2} \times S_X^2 = 1$$

Equating $\Sigma[X - MX]^2/N$ with S_X^2 in accordance with Formula 1-1

$$S_z = \sqrt{S_z^2} = \sqrt{1} = 1$$

Taking the square root

[16] For families of probability distributions other than the normal curves, the numerical values that define the different distributions may not be the mean and standard deviation. The point is that there are families of distributions similar in shape but differing in location (central tendency) and variability.

Since the normal curve is symmetrical, only half of it is tabled and is sufficient for all values of z. This technique may be used with any distribution tabled in terms of z's. *There is no logical connection between the normal curve and z-scores.* The conversion of a group of X-scores to z's will not make the z distribution normal if the X distribution is not.

The second principal use of z-scores is in the comparison of sets of scores measured in different units. For example, if a man has a z-score of $+0.75$ on height and one of -0.50 on weight, we can observe that he appears to be "taller than he is heavy." (This statement may seem difficult to defend, but similar comparisons often arise in psychological investigation.)[17]

If there are known distributions for height and weight, and if the distributions are of near normal-curve shape, then we can use the z's to make statements about the probability of being at least 0.75 standard deviations *above* the mean height and .50 standard deviations *below* the mean weight. It will indeed turn out that this combination is not highly probable for a man selected at random, and a man with these scores can thus be characterized as exceptionally tall and slim—as we said, he is "taller than he is heavy."

But now let's try another example. Suppose we collect teacher's ratings of behavior in a nursery school: specifically, ratings of aggression (X) and dependency (Y). We can tabulate an X distribution and a Y distribution for the 30 children we have handy, but chances are that such distributions will not be perfectly symmetrical and surely not of normal shapes. This may say something about how aggression and dependency are distributed in the population of nursery school kids—that is, not according to the normal curve—or it may imply that teachers ratings do not adequately reflect the (hypothetical) real distribution. Or, to make matters worse, both of these possibilities may be correct.

Our z-score computations permit us to note that Jason is above the average on aggression $(z = 1.25)$ and quite close to it on dependency $(z = +.06)$. It would appear that he is, indeed, "more aggressive than he is dependent." In defense of this assertion, we can state accurately that Jason's aggression score is 1.25 standard deviations above the mean aggression rating in his group (that is what, by definition, $z = 1.25$ means) and that he is .06 (an apparently negligible

---

[17] This discussion involves a controversy among statistically sophisticated psychologists. The controversy rests on (*a*) the distinction between an inherent characteristic and one's measure (X) of it and (*b*) how much the parties feel should be made of the distinction. See further Siegel, S., *Non-parametric Statistics for the Behavioral Sciences*, New York, McGraw-Hill, 1956.

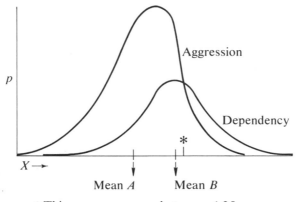

*This score corresponds to $z = 1.25$ on Aggression and $z = .06$ on Dependency.

Figure 1.10 *Distribution of Aggression and Dependency; fabricated data.*

amount) standard deviations above the mean on dependency. The trouble is this: because these scores are ordinal ratings without an objective unit, the standard-deviation unit has no meaning other than to indicate the child's position in his group *vis-à-vis this rating*.

Further, since the distributions of ratings like these are often unknown this "position in the group" cannot be stated in probability terms extending beyond the particular 30 children at hand.

If you encounter comparisons made by means of $z =$ scores, the problem of evaluation centers on whether the comparison *should* be made, not whether it can be; it always can be. It always can be made with a little addition, subtraction, multiplication, or division; procedures that are in themselves neutral with respect to measurement.[18] In Figure 1.10, hypothetical probability curves for aggression ratings and for dependency ratings are presented that make Jason's scores of 1.25 (aggression) and 0.06 (dependency) *equally* probable! These distributions are fabricated and exaggerated; however, they serve to illustrate the point that the sense or nonsense of any comparison of scores is an issue for logical evaluation based on knowledge about the information provided by the scores themselves and by the summary statistics calculated from them.

[18] This statement represents the writer's position on a controversy within the statistics field. For an opposite view see Siegel, S., *Non-parametric statistics for the behavioral sciences.* New York, McGraw-Hill, 1956.

1.5 Summary

A score is a numerical representation of a subject's status with respect to some characteristic. It is assigned to the subject in accord with a measurement rule. This rule determines the appropriateness of any statement made about him with respect to the characteristic. A frequency distribution describes how the individuals who comprise a group are apportioned—distributed—among the possible scores.

From the frequency distribution, summary statistics stating the distribution's central tendency and variability can be computed. The most commonly encountered of these are the mean and standard-deviation for central tendency and variability, respectively. These summary statistics help the reader to visualize the frequency distribution; a process that is, of course, enhanced when graphs are actually presented in the article or book being read.

2

The Logic of
Drawing Conclusions from
Empirical Investigations

An *epistemology* is a theory about how to acquire knowledge. Any such theory involves the formulation of useful questions and the designation of proper procedures for their answer.

Empirical investigation is presently considered by many behavioral scientists the proper procedure for answering those questions about human nature that deal with behavior and experience. An empirical investigation is an attempt to answer questions by making observations that are potentially available to others. The observations may be made in highly structured situations or in less structured ones, and such observations may be direct or indirect.

Thus, for example, looking out the window to see whether or not it is raining is an empirical investigation. Reading today's weather report in the paper while sitting in a windowless room is also an empirical investigation, although the observation is indirect (you read of someone's measurements that have been shown previously to be related to the presence or absence of rain). Sitting in an enclosed, soundproof room and guessing whether or not it is raining is not an empirical investigation, because it is an experience that cannot be shared by others; that is, your guess may not be their guess.

Whether or not empirical investigations constitute *the* proper epistemology for understanding behavior and experience is, of course, subject to debate. In fact, some debate of this kind has been going on for some time. Nevertheless, empirical methods are the primary means of producing data in the behavioral sciences today. This chapter will therefore be devoted to (1) a classification of these methods for your use in identifying them in your readings, (2) an exposition of the logic underlying the use of data in testing an hypothesis, and (3) some suggestions for formulating "conclusions" from these testing procedures and for comparing your conclusions with those found in an author's discussion of his work.

2.1 Research Strategies and How to Identify Them in Use

Most empirical investigations make use of one of three strategies: (*a*) correlational, (*b*) field-descriptive (which often employs correlational techniques), or (*c*) experimental. A large-scale, long-term project may make use of all three strategies at different times.

2.1.1 Correlational Strategy

Any investigation that is aimed at discovering if, and perhaps how much, two or more variables relate, is following a correlational strategy. All members of the group(s) being studied are assigned scores, one per subject for each variable of interest. Various indices to the presence or absence, and degree, of relationship between variables are determined. These details are given in Chapters 3 and 5.

The correlational strategy can be identified in practice by noting that the scores on the various variables are *not* created by the investigator's manipulation. Rather, he measures each individual studied with respect to pre-existing traits. He then relates the scores to each other, hoping to gain insight into how the pre-existing traits, which the scores presumably measure, are related. The conclusions will be more or less correct depending on (1) how valid are his measures of the traits, and (2) how "appropriate" his indices of correlation are to the measures.

2.1.2 Field-descriptive Study

Often in the behavioral sciences, interest centers on complex behavior that cannot easily be observed in the laboratory (see Section 2.1.3) or by means of simple assessment (testing) procedures. For example, it is difficult to glean *all* of the interesting and relevant information about family-interaction patterns, or classroom atmosphere, or language learning from a standardized test or laboratory procedure. In other cases, variables may be potentially available for the better controlled, simpler observation situation of the laboratory, but at the time of the study there may not be sufficient normative information about them to warrant laboratory control over them. For example, we might not want to create artificial political situations in the laboratory until we had a pretty good grasp on the natural operation of the political process of interest. Our fear would be that the laboratory is not relevant to the real process.

In situations such as these, investigators often go into the setting where the behavior(s) of interest naturally occur. Such settings include homes, schools, shops, neighborhood streets, clubs, hospitals, police stations—or almost anyplace *except* the laboratory or testing room.

The aims of a field study may be confined to gathering descriptive information about the individuals studied and the presentation and interpretation of such information in narrative form. A more usual practice is to define measurements of the variables of interest, to assign scores to individuals (or to groups such as families, work crews, scout troops) and to report for each variable summary statistics such as the mean, the standard deviation, and the distribution's shape. This statistical information is usually presented in addition to, rather than instead of, selected narrative records. Both types of information may then be discussed together.

In addition to description of relevant variables, reports of field studies usually contain correlational information about the interrelations among the variables measured. For this reason, many investigators think of such studies as a special case of the correlational strategy. The final interpretive discussion of all the material considers these correlational results in view of the whole picture provided by the narrative records and descriptive statistics. To the extent that correlational information is given, a field-descriptive study includes the correlational strategy. All of the suggestions for interpreting correlational results that are contained in Chapter 3 are pertinent.

2.1.3 Experimental Strategy

The controlled experiment is a very special situation for obtaining observations in which the experimenter controls one or more aspects of the individual's environment or state in an attempt to isolate the variables that are related to a behavior or experience of special interest. For simplicity, let us discuss the two-variable situation first. In the experiment, two variables are again being correlated, but, in addition, one of them (say, temperature) is under the control of the experimenter. The experimenter in fact assigns each of the available individuals to one of several treatments (various temperatures, for example) or arranges for the administration of all treatments to all subjects. Either way, a treatment is equivalent to a score on what we customarily call the *independent variable*. The adjective "independent" refers to the fact that the experimenter assigned each of the subjects to a treatment according to a rule of his own devising; that is, he did it independently of other variables in the study.

The assignment rule is usually that of random assignment subject to certain constraints. Random assignment is such that each individual is assigned independently of the others, and that the probability of being in any one treatment group is the same for all individuals in the supply of subjects (*S*'s). The most usual constraint is that all the treatment groups have the same number of *S*'s. In this instance, after one group is full, the probability of the remaining *S*'s being assigned there becomes zero and the randomness criterion is violated in this one specifiable way.

The measures that are obtained from experimental *S*'s during or immediately after the experimental treatments are measures of the *dependent variable*. The adjective "dependent" here refers to the fact that the designated dependent variable is presumed to depend on the experimental treatment, as well as on other relevant characteristics of the subject. In any case, an experiment is done to see *whether or not* such dependence exists.

Of course various characteristics of subjects other than the independent variable may influence the status of subjects on the dependent variable. An investigator may take steps to have the average values of these other measures equal for all treatment (independent variable) groups. For example, if age is a factor in task performance (dependent variable), then the average age of *S*'s in all treatment groups (independent variable) may be the same. Or the experimenter may take care that the same number of child-*S*'s of each age, 4 through 7, is assigned to every treatment group. These steps are taken to insure that the extraneous (in this context) variable cannot logically be said to account for any differences in dependent-variable status between treatments (independent variable). Such a variable is said to be "controlled" in the experiment, and the constraints on random assignment achieved by this control are deliberate and planned. The main purpose, of course, is to permit the conclusion that such dependent-variable differences that occur between treatments (independent variable) are due to or caused by those treatments and are not attributable to various other variables (the ones that have been controlled).

Good experiments are well controlled, and control of extraneous variation is, indeed, the only way to isolate the operations of independent variables. You should remember that one person's independent variable is another person's control variable. For example, in the study of elementary learning, one investigator may control anxiety level and study the effect of instructions, whereas another may control instruction (by making them identical for all S's) and study changes with situationally induced anxiety states.

2.1.4 The Frustration-Aggression Hypothesis: A Case in Point

This section will illustrate each of the three strategies as they might be employed in the investigation of a single hypothesis. The frustration-aggression hypothesis was derived from Freud's writings; it states that the emotional state of frustration in the individual always leads to a tendency towards aggressive acts by him.[1]

2.1.4.1 Correlational studies. First, suppose this hypothesis was investigated with a correlational strategy. It would be best to note that this strategy could provide us only with information as to whether or not frustration (as measured) and aggression (as measured) are related. Perhaps the degree of such relationship, if present, could be estimated. *But* the correlational strategy does not contain within its procedures any technique to justify the statement that frustration causes aggression. A relationship between them could equally well be cited as evidence that aggression causes one to become frustrated (for example, through guilt feelings) or that frustration and aggression are both caused by the same third variable (for example, parental punishment).

A logical case could be set forth—apart from the statistics—favoring an argument that frustration causes aggression. This argument would have to be evaluated in the same manner in which it was offered, that is, on the basis of the sense it makes. The best safeguard against the perpetuation of incorrect conclusions drawn

---

[1] The classical form of this hypothesis also states that whenever aggressive acts are observed to occur, the acting individuals also must be in an emotional state of frustration.

from correlational studies is to study the same two variables again using different measures. If possible, an experimental strategy should be employed at least once in such a series of researches to check on the causality conclusions.

Two hypothetical examples of the many conceivable correlational investigations relevant to the frustration-aggression hypothesis are as follows: (1) An investigator could ask trained observers to rate children every minute in their natural play situations for both aggression and frustration. There are many variations of this strategy, among which are the number of observers per variable, the scoring procedures for the measures, and the number and duration of observation periods. In any case, each child ends up with an aggression score and a frustration score. A numerical index to the relationship between the two variables could then be computed from the pairs of scores. The choice of what statistic to use for this purpose is discussed in Chapters 3 and 6. (2) The investigator could form groups of subjects on the basis of how much frustration (high or low) they have experienced. The two groups could be formed on the basis of ratings, reports from the *S*'s about how they feel, or judgments made from their life-history files. In any case, all *S*'s could be observed and rated for aggression, as in approach 1, and the distributions for the two groups, especially the means or medians of the aggression scores, could be compared.

The two correlational strategies illustrated are subject to infinite variations due to the kinds of measures employed. The important thing for the reader to note is that both procedures—the single-group approach and the group-comparison approach—*are* correlational. The group differences in approach 2 may be analyzed with statistical tools identical to those used for analyzing differences between groups deliberately formed for an experiment. In this example the groups were formed on the basis of a pre-existing characteristic (frustration level) in the *S*'s. The group comparison in approach 2 is a correlational procedure because both frustration and aggression are measured but neither is regulated by the investigator.

2.1.4.2 Experiment. Next, suppose the frustration-aggression hypothesis were investigated in a controlled experiment. A constant environment could be arranged (for example, a small room with a table and two chairs); within it, the investigator could expose his subjects to prearranged treatments purported to differ in the consequent level of frustration experienced by the subject.

Let's say, for example, that the college student subject is shown a disassembled mechanical gadget and is told that he is to reassemble it and that he can have five dollars if he finishes in five minutes. The experimenter divides his supply of S's into three equal groups at random subject to the restriction that each group contain one-third of the S's. For the high-frustration group he presents fake pieces of a gadget (which simply cannot be assembled), and thus the S's receive neither money nor a sense of completion during the session. The low-frustration subjects are given a disassembled gadget that can be readily assembled, but when they succeed, the experimenter tells them that he can only find four dollars in his cash box, and thus they receive less money than expected for completing their task. The no-frustration subjects assemble the gadget readily and are given the five dollars. The subjects' behavior just after leaving the experimenter (their remarks to other S's in a waiting room) might form a basis for measuring the dependent variable.

In this example, the experimenter has decided which subjects are to have which scores (high, low, no frustration) on the independent variable. If he is careful he will take steps to insure that his three groups of subjects are comparable with respect to their predisposition to aggressive reactions to frustration. He may do this by adding additional constraints to the assignment procedure to insure that the group means, standard deviations, and distribution shapes are similar for some measure—presumably of a predisposition to aggression. Alternatively, he may include the predisposition characteristic as a second independent variable (see Chapter 4), or he may take all of his subjects from a group of people known to be very similar to each other with respect to the predisposition factor. Any of these arrangements would constitute a *control procedure*, so named because the experimenter gains control (in the everyday sense) over variation in the predisposition characteristic.

It is likely that you will encounter the terms "control group" and "control treatment." These refer to groups of subjects that are subjected to all aspects of the environment and incidental stimulation attendant to those in the experimental group(s) but who receive no specifically induced treatment. In our example, the no-frustration group is a control group; it has the same instructions, task, and reward expectation as the other two groups, but frustration is not experimentally induced.

Here, it is not predisposition that is being controlled; it is any effects of the treatments *other* than the introduction of frustration. This control can be effected by having a separate no-frustration group as in the example or by administering all treatments, including the control treatment, to each S and comparing his behavior under it with his behavior under experimental treatment(s).

2.1.4.3 Field-descriptive studies. The frustration-aggression hypothesis could be investigated by observing children in everyday settings. Observers could be asked to watch the children at play, in school, at home, out shopping, in the dentist's office, or almost anywhere. Taped or typed records are often made of such observations. These records are then scored for various measures; in this example, frustration and aggression. Several different kinds of frustrating situations and several different kinds of aggressive behavior may be scored; the two measures may then be subjected to a correlational analysis.

2.1.4.4 Summary. In all three types of study, two (or more) variables are measured and some conclusions can be made about their systematic relationship. In any experiment, the independent variable is measured by definition—by deliberate assignment to treatments that are taken as measures (or at least categories) of the independent variable. The other variable is assessed in an environment that is the same for all subjects except for those aspects of it that are part of the experimental treatment. In a correlational study, the two variables are measured by the responses of S's to available stimuli. Frequently (but not always), the stimuli are standardized, as would be personality test items, pictures to which one is to respond, or verbal instructions. Field studies are similar to correlational work, but they permit greater flexibility of environmental and time factors; this flexibility is, of course, the opposite of experimental control. The three types of study we have defined differ with regard to the degree of investigator control that is possible. The reader's job requires, often, comparing several investigations of the same hypothesis (say, the frustration-aggression hypothesis) done under different degrees of investigator control. Fawl (1963), for example, reports "no relation" between frustration and aggression in his field study of children in a Midwest community. Sears et al. (1957) found moderate relationships between their several measures of parental frustration and children's aggression. McClelland and Apicella (1945) found S's who received insults from an experimenter (frustration) while performing a manual task were verbally aggressive towards the experimenter, whereas control S's were not. If this general pattern—that is, that the frustration-aggression hypothesis seems to have stronger support in the more controlled situations—was repeatedly noted, the hypothesis should be restricted. Also, perhaps the definitions employed by different investigators of frustration and of aggression should be evaluated with a view to gaining further insight into the nature of these two variables and their relationship.

2.2 The Logic of Testing a Hypothesis

The purpose of some field-descriptive studies is only to gather previously unavailable descriptive information; such studies depend very little on previous expectations. It is more usual for an investigation (of any type) to be undertaken for the purpose of discovering the merits, if any, of some prior expectations. In the context of designing experiments, defining measuring devices, and planning data analyses, these expectations are expressed formally, and perhaps mathematically, and are called "hypotheses." Here are three examples of hypotheses: (1) An experimental group will persist at a task for less time than will a control group. (2) Small girls and boys will differ systematically in their verbal test scores—girls will tend to have higher ones. (3) Punitiveness in parents will be directly related to aggressiveness in children.

A mathematical statement of the first hypothesis would be expressed in terms of the expected difference between two means; that is, for X = the time spent on a task, $\bar{X}_e > \bar{X}_c$ or $\bar{X}_e - \bar{X}_c > 0$, where $\bar{X}_e$ stands for the mean of the experimental group and $\bar{X}_c$ for the mean of the control group. Similarly, the second hypothesis would state that $\bar{X}_g > \bar{X}_b$, or $\bar{X}_g - \bar{X}_b > 0$, where $\bar{X}_g$ and $\bar{X}_b$ represent girls' and boys' means. The third hypothesis would be expressed in terms of ρ, a population measure of one degree of relationship (see Chapter 6). It would say simply: $\rho_{xy} > 0$, where x = parental punitiveness score, y = child's aggression score, and ρ_{xy} is the sample value corresponding to ρ. All of these hypotheses can be tested according to a classical logic that depends on probability.

2.2.2 Probability Terms

Recall from Chapter 1 that the probability of a score is defined as its *theoretical relative frequency*. In symbols

$$p(X) = f(X)/n_{\text{pop.}} \tag{2-1}$$

where $p(X)$ is the symbol for the probability of x; $f(X)$ is the symbol for the frequency of x; N_{pop} is the number of scores in the entire population. This definition can be extended to events other than scores, such as means or differences between means; standard-deviations, correlation indices, or other statistics. In these instances, $N_{\text{pop.}}$ is the number of means (or other statistics) in some population of means (or other statistics) and X in Equation 2-1 would be replaced by a symbol such as $\bar{X}$ representing the statistic. In general, the symbol E stands for an unspecified event. Thus, $p(E) = f_{(E)}/N_{pop.}$; $N_{pop.}$ is often, but not always, equal to infinity. This is explained further in the next section.

2.2.2 Population and Sample

2.2.2.1 Representing the population. In any investigation, there is a specific group of subjects being studied. The members of this group provide the scores, the means, medians, frequency distributions, standard deviations, and any other numerical information used by the investigator in offering his interpretations and conclusions. This immediately poses a question for the reader: are these means, medians, and so forth, relevant to anybody else? More particularly, can the conclusions offered by the investigator, or by the reader, be applied to some class of individuals larger than the group actually studied? For example, a teacher is interested in junior high school pupils *in general*, not just in 37 eighth graders from P.S. 5001 in New York. An author of an introductory text is interested in formulating as general a statement as he can about the effect of anxiety on physical performance, not just on anxiety in 28 football players from Western State University. The clinical psychologist wants to know if some form of group therapy is a good bet for most couples with marital problems (in his practice) not just for 15 couples at the Lovers' Lane Marriage Counseling Clinic.

To suit your own purposes, you need to know two things about the group being studied. First, what larger class of individuals does this specific group represent? Second, is this larger class related to the class of individuals about whom I wish to draw conclusions? The larger class of individuals represented by the sample is called the *population.* The group studied is called the *sample,* and it is assumed that the sample is representative of the population.

What does "representative" mean? A sample of N individuals represents a larger population to the extent that the various characteristics of members of the population are distributed in the sample proportionately to their distribution in the population. For example, if the population of students at a college contains 65 percent men and 35 percent women, then so would any representative sample of the student population. Similarly, the ages, college majors, marital statuses, numbers of children, fathers' occupations, and other characteristics of the individuals in the population should be found in the sample in the same proportion as they are in the population.

In practice, it is not possible to have any sample perfectly representative of any known population. For one thing, there is a virtually infinite number of characteristics of individuals in populations. To draw samples of any manageable size that proportionately represent some population on all characteristics is simply an impossibility. Furthermore, the practical problem is the reverse of the sampling theory problem. In practice we ask, given these N people, who do they represent? Samples are usually drawn by one of two strategies, each of which has something to recommend it, but neither of which produces perfectly representative samples.

2.2.2.2 Methods of sampling. The first method of drawing a sample from a known population is that of *random sampling*, whereby members of the population are chosen for membership in the sample in such a way that all possible samples of N individuals have identically equal probabilities of occurrence. Samples of size N are formed by taking individuals one at a time from the population *without replacement.*

Thus, for example, if we have a population of Huey, Dewey, and Louie, there are three possible samples of size $N = 2$. These are {Huey, Dewey}, {Huey, Louie}, and {Dewey, Louie}. Each sample would occur one-third of the time in the long run; all three have probabilities of occurrence equal to one-third and are therefore *random samples.*

The second method of selecting a representative sample is known as *stratified sampling*. This procedure may be conceptualized most easily by extending the definition of random sample. The investigator decides in advance how important it is that various characteristics of individuals be proportionately represented in the sample. He then divides the larger population into subpopulations—or *strata*— according to the presence of different combinations of important characteristics. For example, if he wants his sample exactly representative of the population for sex and age, he divides his population into males and females and into age groups. He then has a number of age-sex strata equal to the number of age groups times two (the number of sexes). He draws *at random* from each *stratum* the number of individuals necessary to insure that the sample has the same proportion of such individuals as does the population. For example, if the population is all the grade school students in his city, he would have the seven ages 5 through 11 and the two sexes, making 14 strata. If 7 percent of the city schools' pupils were 10-year-old girls and he desired a sample of size 200 from the population, then such a sample would necessarily include 14 (7 percent) 10-year-old girls. These girls would be selected *at random* from the relevant stratum of the population; that is, the 10-year-old girls. All other strata would similarly be sampled at random so that the final sample of 200 would truly represent the children in the city schools (population), at least for numbers of boys and girls and for numbers from each of seven age groups.

It should be noted that stratified sampling is not *necessarily* superior to simple random sampling from the population. The logic of random sampling is such that if the population is infinite and the sample size is very large, the sample will indeed be approximately representative of the population with regard to all of the attributes of the individuals in the population. One usually encounters stratified sampling when the population is finite (200,000 pupils in the city schools) and the investigator wants to make inferences about—or at least to describe—the attributes of the known finite population. The stratified procedure insures representation of this population in the sample.

2.2.2.3 A reader problem: Given a "random" sample, what population has been represented? Most readers of the literature have in mind some population of individuals to which they would like to generalize the findings they encounter. For example, one may be interested in the development of personal values in the undergraduate college years. Any given study of college students' values will report results based on *some* students who attend a particular college or colleges. Suppose, for example, that 70 percent of 245 undergraduate students at Badlands State College state to an interviewer that they are somewhat favorable to complete societal permissiveness for sexual activities between consenting persons. We will see later that, if sampling has been properly carried out, this 70 percent can be considered the best estimate of the population percent. But here is an important question: *what population?* Is it (*a*) all students at Badlands State? (*b*) all students at four-year colleges in the same state or region as Badlands? (*c*) all undergraduates in the United States regardless of type of institution? (*d*) all persons in the 17- to 22-year-old age ranges? (*e*) all middle-class persons in that age range? (*f*) all college students from student bodies equal in size to Badlands'? A case could be made that each of these is *the* population. But some cases would be better than others. The dilemma you face as a reader is that the sample members have already been chosen by someone. You, will have to make an educated guess about who they represent and what phrase would best characterize the population.

The investigator may name the population he believes is being sampled. In fact, he may demonstrate that his subjects were indeed selected at random from all enrolled students at Badlands State. If so, surely "Badlands State student body" would be the most conservative description of *the* population. If either the investigator or the reader wants to generalize more broadly, he will have to demonstrate with figures or logic that Badlands State students are representative of some larger group of individuals, such as group *b* above. The more usual solution is to claim that the conclusions drawn from the 245 Badlands State students apply to some *abstract and hypothetical* population known as something like "young people typified by Badlands State students."

The purposes for which exact identification of the population is of special importance are (1) practical application of the findings or (2) comparing findings from the same general technique applied to two different samples that may or may not represent the same population.

2.2.3 The Problem of Estimation

The findings of an investigation will usually be presented and discussed in terms of some of the descriptive statistics described in Chapter 1. For example, one may read a percent, a mean, a standard deviation, a difference between means for two groups. Indeed, one may read tables that contain several such statistics. *These statistics are computed from the sample data,* which are the scores for all individuals sampled; statements are, however, made *about the population.* The statistics are used as estimates of the corresponding values, which are called *parameters* in the population. This is the crux of the matter—of what inference is all about. The obvious question is "How good an estimate is this mean (standard devision, or whatever)?" The answer to this question is best expressed in terms of something known as a *sampling distribution.*

2.2.4 Sampling Distributions

To begin with, the X-scores are estimates of individuals' real statuses on some trait or interest. The mean, $\bar{X}$, of the scores is an *estimate* of the central tendency of all of the X-scores that belong to individuals in the entire population, including the sample members. The population is larger than the sample and, therefore, permits many (perhaps an infinity) of samples to be made. The sample at hand in one investigation is, of course, only one of these. The sample mean, $\bar{X}$, is thus only one of many possible estimates of the population mean, μ.

The answer to our prior question, "How good an estimate is $\bar{X}$?" depends on the nature of a special kind of probability distribution (review Section 1.0) known as *the sampling distribution of the mean.*

The reader of this book is not likely to deal directly with such distributions. He will, however, need to know something about the sampling distributions most commonly used in order to form conclusions based on statistical analyses that derive from them.

Let us consider the definition of a sampling distribution by reference specifically to the sampling distribution of the mean. Recall that a sampling distribution is a probability distribution; this means it is by definition a distribution of theoretical relative frequencies. Now imagine that we drew a random sample of size N (say, 50) from some identified population. Next, suppose that we recorded the mean, $\bar{X}_1$, of the N scores and *returned* them to the population. Then suppose we drew another N scores and recorded their mean, $\bar{X}_2$, and returned them, and so on. *The returning is important;* it insures that, at the time each sample is constituted, all scores in the population of scores have an equal chance of being represented. If we retained each sample instead of returning it, we would remove all chances that any score in sample 1 could also be in sample 2. This means that sample 2 would be a random sample of the original population *minus sample 1*. When we replace sample 1 into the population, we insure that sample 2 is just as much a random sample of the population as was sample 1. So are all subsequent samples.

Here is a familiar classroom example of sampling. The instructor produces a large coffee can containing 500 slips of paper, each with one of the numbers "0" to "9" on it—this is the population. He shakes the can and draws 10 of them; these 10 he calls sample 1. He returns them to the can, shakes the can again, and draws another 10, which he calls sample 2. Sample 2 may or may not contain any slip which was in sample 1. This is called sampling "with replacement" and the replacement is necessary for the claim that each successive sample is a random selection from the *original* population. To make this even more concrete, let us note that "without replacement" there would be 50 (500 ÷ 10) possible samples and that "with replacement" there would be several billion possible samples (the number of combinations of 10 things that can be taken from 500 things).

Each sample of size N will consist of N scores, and these N scores will have a mean. If we draw a large number, N_s, of samples we will then have this same number, N_s, of means. In the coffee can example, there will be means ranging from 0 to 9, since this is the possible range of scores. It would be possible, though not probable, for a sample to consist of all 0's (that is, $\bar{X} = 0$) or all 9's (that is, $\bar{X} = 9$). In any case, we could construct a frequency distribution for any number of means, based on a fixed sample size, say 10. It might look like the first three columns of Table 2.1. There are nine intervals,

Table 2.1   *Frequency distribution of 100 means from samples of size 10*

Interval of means	Midpoint	f	Relative f
8.00–8.99	8.5	0	0.00
7.00–7.99	7.5	1	0.01
6.00–6.99	6.5	5	0.05
5.00–5.99	5.5	21	0.21
4.00–4.99	4.5	50	0.50
3.00–3.99	3.5	17	0.17
2.00–2.99	2.5	6	0.06
1.00–1.99	1.5	0	0.00
0.00–0.99	0.5	0	0.00
		100	1.00

Mean = 4.55   Standard deviation (SD) = .953

corresponding midpoints, and corresponding frequencies. If we imagine extending the quantity of samples (each sample containing 10 scores) to larger and larger numbers, approaching the billions possible, then the relative frequencies (listed in column 4 of Table 2.1) become, *by definition*, the probabilities of the corresponding means at the midpoints of the intervals of column 2 of Table 2.1. Thus, columns 2 and 4 of Table 2.1 can together be considered an approximation of a probability distribution for means of samples of size $N = 10$ of scores ranging from 0 to 9. Of course, greater accuracy could be obtained by recording individual means instead of intervals of means.

A probability distribution for means, such as that described above, is known as a *sampling distribution of the mean*. Like any other probability distribution, this sampling distribution of the mean has its own mean and its own standard deviation. These are known and symbolized as μ, *expected value of the mean*, and $\sigma_{\bar{X}}$, *standard deviation of the mean*. Any descriptive statistic will have its own sampling distribution with associated expected value (mean) and standard deviation. Note at the bottom of Table 2.1 that the mean of the sampling distribution is 4.55 and the standard deviation is .953. More generally, when the mean, μ, and standard *deviation*, σ, of the distribution of scores are known, it can be shown that μ is the mean of the means that the standard deviation of the mean, $\sigma_{\bar{X}}$, is $\sigma/\sqrt{N}$. When the standard deviation of the X-scores is unknown, it may be estimated by the S computed from sample data where S is the square root of the variance as defined by Equation 1-1. In this case the standard deviation of the mean may be estimated by a quantity known as the standard error of the mean, $S/\sqrt{N}$. Similarly, when μ is unknown, it may be estimated by $\bar{X}$, the sample mean. Note here that the expected value for the X-scores in a sample is the same as the expected value for the sample mean, that is, μ.

The sampling distribution of the mean has, of course, a shape. *The central limit theorem tells us that, for known population parameters μ and σ and at least moderately large N, the sampling distribution of the mean will have normal shape.*

Any sampling distribution has some shape and those of the familiar descriptive statistics are known and form the basis of evaluating estimates and hypotheses about them. We are now ready for our question, "How good an estimate is this?" and the related one, "Is this hypothesis true?" Let us proceed to the first.

2.2.5 Estimation

The answer to this question is often discussed in basic texts (see, for example, Hayes, 1963, pp. 196 ff.) by describing properties that should be possessed by any statistic that is proposed as an estimate of a parameter. The four principal properties can be reviewed in an intuitive manner by describing them as they apply to the same mean, $\bar{X}$, as estimate of the population mean, μ:

(a) If we average the means, $\bar{X}_1, \bar{X}_2, \ldots, \bar{X}_k$, of all possible samples (k large, possibly infinite), this "mean of the means" should equal μ. This suggests that overestimates in some samples are cancelled out by underestimates in others. This is the property of *unbiasedness*. We might note here—since it comes up later—that the sample standard deviation, S, is not an unbiased estimate of σ, the population standard deviation. A corrected estimate, $\hat{S} = \sqrt{N/N-1}\,S$, is often used, since the degree to which it is biased is much less than that of S. The standard deviation that is given in descriptive summaries of an investigator's data may be calculated from either the formula for S or that for $\hat{S}$. Unfortunately, there is no standard practice in the journals as of this writing.

(b) The absolute value of the difference between $\bar{X}$ and μ, that is, $|\bar{X} - \mu|$, should tend to become progressively smaller as the size of the sample increases. Thus one ought, in principle, to be able to make that difference as minute as desired by making N as large as possible. Thus if we drew a sample so large as to take every member of the population, then the difference, $|\bar{X} - \mu|$, would be zero. This property is called *consistency*, and is related to what is known as the "law of large numbers."

(*c*) $\bar{X}$ should have a standard error smaller than that of other candidates for estimation of μ. This means, among other things, that the sampling distribution of $\bar{X}$ would appear visually to be narrower than that of other candidates. This property is known as the *efficiency* of the estimator $\bar{X}$ relative to other candidates. The mean is efficient relative to the median, given certain assumptions about the distribution of the X scores.

(*d*) $\bar{X}$ should have a calculation formula that uses all of the available X scores (all the sample data) so that no improvement would result from use of additional sample information; this property is called *sufficiency*.

Any statistic—not just the mean—can be evaluated for these abstract properties. It is not important for you to be able to make these evaluations; however, you do need to know why they are used. In other words, you need to know that what may appear to be arbitrary directives that a given formula be used for an estimator of a property (mean for central tendency in one situation and median in another) are actually based on logical criteria of what makes a *good* estimate.

When it is less important to know the exact value of a parameter than it is to know with high assurance what ball park the parameter is in, then *interval estimates* are used. Techniques for determining what are known as *confidence intervals* are available in any basic text. The K percent confidence interval for μ, say, consists in two numbers, a lower and an upper limit. The probability is $K/100$ that the technique will produce two numbers that include μ between them. A confidence interval will be expressed by some such statement as: "The 95 percent confidence interval for μ is 8.62 to 9.79."

You will rarely note any reference to estimation as such in research reports; you will see tables of means, standard deviations, and other descriptive information. Most often, this information will be evaluated and interpreted in terms of *hypotheses* about what the parameter values corresponding to these statistics might actually be in the population. The two topics estimation and hypothesis testing are discussed separately in basic texts for the purpose of clarifying their *logical* distinction. In practice, however, the two questions, "Is this a good estimate?" and "Is this hypothesis true?" cannot be answered independently of each other. We have to answer the second on the assumption that we have a good estimate of the relevant parameter. The reader must live with the fact that the assumptions that are involved in deciding that an estimate is or is not good are involved also in testing hypotheses regarding the value of the parameter.

2.2.6 Testing a Hypothesis: I. The Basic
Logic of All Tests: Single Mean Example

Let us explore the logic of hypothesis tests in general in the context of a very simple—but realistic—example. Suppose you are planning a study that will evaluate the effectiveness of social reinforcements (experimenter approval) on the learning of a standard laboratory task involving motor skills and eye-hand coordination. You have a sample, say 50 subjects, to be divided into two groups—one to be reinforced, the other not. You wish to be sure that your 50 subjects do not differ from normal in their initial ability on the task in order that there will be some room for improvement. Specifically, you want there to be some differential improvement in the two groups. An investigator with this problem might pretest his 50 subjects on the task and compare their performance with some known or hypothesized population mean, μ.

Now, how would μ be known? In the example, the standard-apparatus manual may report the number of tries (trials) it took for mastery for a large number of subjects (say 20,000) who were tested in its development. The investigator then rephrases his problem as follows: If these 20,000 tested individuals constitute the population, could my 50 potential subjects be a random sample from that population? If the answer is yes, the investigator will proceed to divide them into two groups and use them without further ado. If the answer is no, he may seek different subjects, or he may use this information to adjust his analyses and the interpretation of his results. In any case, his problem can be phrased as that of a decision to *accept* or *reject* the hypothesis that the mean of the population from which his subjects came is identical with that listed for the 20,000 tested subjects.

Let us review the appropriate symbols:

μ = Population mean—in this case the mean for the population from which the 50 subjects came.

$\bar{X}$ = Mean of the sample of 50 subjects.

X = The variable whose mean is being questioned. In this instance, X = trials for a subject to master the task. A higher score implies poorer performance (slow learning).

N = Number of subjects in the sample. Here, $N = 50$.

σ_x = Standard deviation of the X scores in the population (known).

Let us further suppose that the apparatus manual informs us that for the test population of 20,000 subjects, $\mu = 6.19$ and $\sigma = 1.10$.

Then we can state the hypothesis (H) being tested as, H: $\mu = 6.19$. In other words, we hypothesize that our 50 subjects come from a population with $\mu = 6.19$.

A procedure used to decide whether or not the hypothesis, H, is plausible is called a test of H. Our only pertinent data produce a single mean, $\bar{X}$, and we must decide whether or not it is reasonable to conclude that $\bar{X}$ is one of the means in a sampling distribution centered at $\mu = 6.19$. This procedure requires us to know the mean, standard deviation, and shape of the distribution of means drawn from a population such as the 20,000 tested subjects. The classical steps of logic are as follows:

Step 1: Recall that the sampling distribution of means always has as its mean, $\mu_{\bar{X}}$ equal to μ, the population mean for X. In this case $\mu_{\bar{X}} = \mu = 6.19$, *if H is true.*

Step 2: Assume the nature of the shape of the sampling distribution of means. In this case $N = 50$ is large enough to warrant invocation of the central limit theorem. Thus, we assume the shape to be that of the normal curve.

Step 3: Assume the value of σ of the sampling distribution. In this case σ_X is known to be 1.10. The sampling distribution has a standard deviation equal to $1.10/\sqrt{50} = 1.10/7.07 = .141$.

Now, let us pause to take full note that each of steps 1 through 3 makes an *assumption which may or may not be true.* Step 1 assumes the hypothesis to be true and is the only one that will be directly questioned or tested in the procedure outlined here; this is important. When you read that a hypothesis is being rejected, bear in mind that there were also some other assumptions required for the test and that it would have been permissible (although perhaps not sensible) to reject one or more of these other assumptions instead of, or in addition to, the hypothesis.

(As so often before in this book, I am emphasizing the importance of reasoned judgment in interpreting statistical analyses: the choices about acceptance *or* rejection—and of what to accept or reject—are ultimately matters of reasoned personal judgment. The reason the whole procedure does not reduce to an idiosyncratic guessing game is that we agree to follow certain specified steps in forming our judgment. We assume that others reading our empirical work will follow similar steps in forming their judgments of it or, alternatively, will explicitly state that their judgments differ because they dispute our logic. Let us continue, then, in an exposition of the customary logic.)

Step 4 is essentially the comparison of the observed mean for 50 subjects with the hypothesized population mean. Suppose $\bar{X}$ is calculated, and we find

$$\bar{X} = 6.47$$

The first assumption, the hypothesis, states that $\mu = 6.19$, so the difference between $\bar{X}$ and μ is $\bar{X} - \mu = .28$.

In step 5, we ask the question, "Is this difference large?" and seek an answer in terms of probability. Steps 1 through 3 allow us to assert that the standard score,[2] $(\bar{X} - \mu)/\sigma_{\bar{X}}$, will follow the unit normal-curve distribution (see Chapter 1, Section 1.2.3). To be specific, the sampling distribution of means, $\bar{X}$, centering at $\mu_{\bar{X}} = 6.19$ is normal (step 2) and, since μ and σ are known constants (steps 1 and 3), the standardized scores formed from these means, $\bar{X}$, will be normally distributed. Multiplication, division, addition, and subtraction of constants does not alter the basic probability distribution of a variable. Now we are in a position to ask *where* in the normal probability distribution of means, $\bar{X}$, does the observed mean, 6.47, fall. To answer, we calculate

$$z - \frac{\bar{X} - \mu_{\bar{X}}}{\sigma_{\bar{X}}} = \frac{6.47 - 6.19}{.141} = \frac{.280}{.141} = 1.98$$

We then proceed to use a standard table of normally distributed z scores to find out what the probability is that a z score of this absolute magnitude could have resulted from one of the many sample means, $\bar{X}$, in the sampling distribution of means centering at the hypothesized value of 6.19. Since our investigator is interested in whether his subjects are different in ability—whether less or more capable than the population—the proper question now becomes, "How likely, *given all of the assumptions*, is it that a z score larger than the calculated value 1.98—or smaller than -1.98 (that is, at least 1.98 in absolute magnitude)—could have arisen from a distribution of z scores based on means in the sampling distribution centering at $\mu_{\bar{X}} = \mu = 6.19$?" The investigator, in referring to his table of the normal curve (as depicted in Figure 2.1), will find that a z value *at least* as large in absolute magnitude would occur with probability .048. This tabulated value equals the proportion of area beyond $z = 1.98$ and $z = -1.98$ on the depicted curve.

In step 6, we make a decision about the hypothesis $\mu = 6.19$. This decision is clearly based considerations of probability. Stated simply,

[2] In the context of hypothesis testing, the standardized score corresponding to a mean, $\bar{X}$, of N scores is sometimes called a "critical ratio," or C.R.

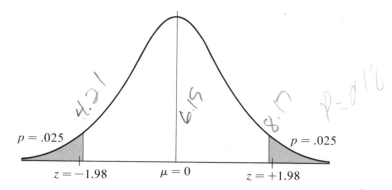

Figure 2.1 *Normal curve showing shaded rejection region for $p = 0.05$ and indicating the z values obtained in the example of Section 2.26.*

any assumption that makes the observed data improbable is rejected. If the data ($\bar{X} = 6.47$) are improbable given the assumptions, then one of them—namely, the hypothesis—is *rejected*. In our example, the investigator would conclude that his 50 subjects would *probably* not constitute a random sample from the hypothetical population with $\mu = 6.19$, and that, therefore, they must be a random sample from some population with some other mean, $\mu \neq 6.19$.

This concludes the step-by-step test of a simple hypothesis about a single mean. Let us briefly summarize the six steps.

1. Form a hypothesis that the value of the population mean, μ, is some specific number. Base this on previous suggestive data, a standard known population, mystic insight—anything you like.

2. State the shape of the sampling distribution of the mean. When the central limit theorem applies, this is normal. (See the next section for an alternative when the central limit theorem does not work.)

3. Determine the value of the standard deviation of the sampling distribution of the mean, $\sigma_{\bar{X}} = \sigma/\sqrt{N}$.

4. Determine the sample mean, $\bar{X}$, and subtract the hypothesized value of μ from it.

5. Determine the value $z = \dfrac{\bar{X} - \mu}{\sigma_{\bar{X}}}$ and refer to a table of the sampling distribution of $\bar{X}$ to determine how probable a value of z having an absolute magnitude that large or larger would be.

6. If the probability is small, decide the hypothesis is not true and *reject it*. Otherwise, accept it.

The probability (p) that is considered to be small is arbitrary, but there are some guidelines (discussed below) that suggest certain values. For many purposes, $p = .05$ is small enough.

When an hypothesis is rejected in this manner, the investigator will usually state in his article that the sample mean was found to differ *significantly* from the hypothesized mean *at the* .05 (.01, .02, whatever) level. He is saying that his observed $\bar{X}$ is large enough to have probability *less than* .05 (.01, .02, or whatever), given that the hypothesis about μ was true. (Such an arbitrary probability value is called the *significance* level.) He therefore *decided* that the hypothesis about μ could not be true. This implies that his $\bar{X}$ was a member of some other sampling distribution of means centering at some other value of μ, which value is unknown.

In other words, he could not question the actual data in hand, so he questioned the assumptions; as a result, he abandoned the hypothesis.

2.2.7 Testing a Hypothesis: II. Single Mean Example with Population σ Unknown

Let us reconsider our six steps in Section 2.2.6., assuming this time that no value is available for the population standard deviation, σ_X, and that we have only 25 subjects. Suppose our population mean, μ, is again 6.19. The sample data provide $\bar{X} = 6.47$ (as before) and $S = 1.05$ (S is an estimate of σ_X.) In this instance, the assumption in step 5 that the ratio $(\bar{X} - \mu)/\sigma_{\bar{X}}$ will follow the same distribution as $\bar{X}$ does not apply. The ratio in this example, called t, is $(\bar{X} - \mu)/\hat{S}_{\bar{X}}$ in which the denominator, $\hat{S}_{\bar{X}}$, is not a known constant, but an estimate that has its own sampling distribution. $\hat{S}_{\bar{X}}$ is defined as $\hat{S} = (\sqrt{N/N - 1})\, S$.

This $t = (\bar{X} - \mu)/\hat{S}_{\bar{X}}$ is the ratio of two random variables and has its own sampling distribution, the t-distribution. Actually, there is a family of t-distributions, one for each number of *degrees of freedom* used in estimating the denominator of the t-ratio; the degrees of freedom are essentially the number of scores that are free to vary when the formula for calculating the denominator is imposed. The denominator for a single mean is an estimate of the standard deviation of the mean, and it has $N - 1$ degrees of freedom.

Thus we make some changes as we follow the previous six steps:

1. No change. Hypothesis states $\mu = 6.19$.
2. Assume the t-distribution with $N - 1$ or 24 degrees of freedom to be the shape of the sampling distribution of the $t = (\bar{X} - \mu)/\hat{S}_{\bar{X}}$.

This assumption can be true only when the scores $X_1, X_2 \ldots X_N$ are themselves normally distributed; thus this assumption implies step 2a.

2a. Assume the scores $X_1, X_2 \ldots X_N$ to be normally distributed.

3. Estimate σ by $\widehat{S}$, a minimally biased estimator equal to $\widehat{S} = (\sqrt{N/N-1})\, S$, where S is the sample standard deviation, 1.05. Estimate $\sigma_{\bar{X}}$ as $\widehat{S}/\sqrt{N}$.

4. Find $\bar{X} - \mu$.

5. Determine $t = (\bar{X} - \mu)/\widehat{S}$ and refer to a table of t for 24 degrees of freedom (abbreviated $df = 24$) to determine how probable a value of t at least this large (in absolute value) would be.

6. Make a decision about H as before.

In summary, the t-distribution could be cited by an investigator in any test of an hypothesis about a single mean where the population sigma is unknown. If N is very large, the use of the normal distribution has some justification in that (*a*) the central limit theorem would justify the assumption of normal sampling distribution for $\bar{X}$ and (*b*) the large N would suggest that $\widehat{S}$ is a good estimate of σ. The t-test, as it is called, is the more conservative. By that, it is meant that for moderate N it will give slightly larger probabilities than the normal curve in the extreme portions (see Figure 2.2). It is readily seen that for a given significance level, it will be more difficult to reject H using the t-distribution than using the normal curve.

Let us return to our example from Section 2.2.6., in which the hypothesis is that $\mu = 6.19$, and the actual data provide $\bar{X} = 6.47$ and $S = 1.05$. The appropriate estimate of $\sigma_{\bar{X}}$ would be $S =$

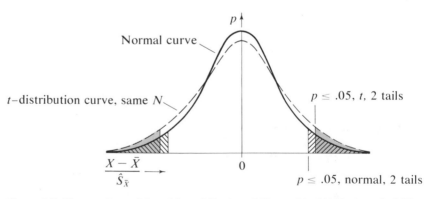

Figure 2.2 *Comparison of the t (dotted line) and Normal (solid line) probability distributions of sample means for N small.*

$(\sqrt{N/N-1})\,S = (\sqrt{25/24})(1.05) = 1.071$. Thus the *t*-ratio would be

$$t = \frac{6.47 - 6.19}{1.071/\sqrt{25}} = \frac{.28}{.2142} = 1.307 \simeq 1.31$$

There would be 24 degrees of freedom. The investigator would discover from a table including both tails of the distribution that a *t*-ratio at least as large in absolute value as 1.31 would occur more than a proportion .05 of the time. If he had selected the .05 level of significance, he would *accept* the hypothesis. Namely, he would decide that a *t*-ratio calculated with X as 6.47 could belong to a (*t*-shaped) sampling distribution of mean centered at zero, the value of the *t*-ratio calculated with $X = 6.19$. Note that (*a*) with $N = 50$ and σ known, the hypothesis H: $\mu = 6.19$ was rejected and that (*b*) with $N = 25$ and σ estimated, it was accepted. If the investigator had erroneously assumed a normal distribution for these ratios, he would have rejected H_0 when, in fact, p was greater than 0.05.

2.2.8 Note on the Directionality of Hypotheses

The reader will encounter references to "one-tailed" versus "two-tailed" tests of hypotheses, or, sometimes, to "one-tailed" or "two-tailed" probabilities. The investigators are making use of the conventions regarding how hypotheses and their alternative hypotheses should be stated. In order to develop a decision procedure for accepting or rejecting a single hypothesis such as H: $\mu = 6.19$, all other possibilities about μ must be included in what we shall call the *alternative hypothesis*. To keep them straight, the hypothesis subject to test is called the *null hypothesis* and symbolized H_0. The alternative state of affairs is called H_1.

The important points are that there must be only two hypotheses, and they must be mutually exclusive and exhaustive. This permits rejection of one to imply acceptance of the other. There are *three* logical possibilities regarding a hypothesized value for μ. Consider our example; either $\mu = 6.19$, $\mu > 6.19$, or $\mu < 6.19$. To reduce the possibilities to two, the convention is to state H_0 either as $\mu = 6.19$ (two-tailed) or as $\mu \leq 6.19$ (one-tailed). The first version requires H_1 to be $\mu \neq 6.19$; this produces the requisite two mutually exclusive and exhaustive alternatives. This version of H_1 is useful when there is no reason to expect that μ might differ from 6.19 in one direction rather than the other. The second version of H_0 permits H_1 to be stated as $\mu > 6.19$; this is a reasonable choice if there is reason to expect (a research hypothesis) μ to exceed the μ of the standardized population. In our example, this expectation that $\mu > 6.19$ would mean we believe our sample comes from a population that takes longer to do the task than the standard one.

The term "two-tailed" refers to the fact that a test of the first version of H_0 requires that differences from the hypothesized mean *in both directions* be considered in terms of probability. Graphically, and in tables based on graphs, such a test may involve both tails of a distribution. The term "one-tailed" refers to a situation in which differences in only one direction (say "greater than") the mean are considered. This requires only one tail (or its tabled equivalent) to be checked. In some cases, a two-alternative hypothesis may require only one tail of an appropriate distribution (such as the F-distribution in Chapter 4). Because of this, the term "two-sided" test is probably more precise; however, the "one-tailed" and "two-tailed" terminology is commonly seen in journals of the behavioral sciences. Acceptance of H_0 under the two-tailed strategy is usually discussed as if it had been accepted that $\mu = 6.19$ (or whatever); actually, it has been accepted that $\mu \leq 6.19$. The choice for discussion between $\mu = 6.19$ and $\mu < 6.19$ is usually based on the size of the sample mean $\bar{X}$. Strictly speaking, if H_0 is accepted, classical logic calls for no further inferences, formal or informal, to be made about μ.

2.2.9 Being Wrong

Since hypothesis tests are based on probabilities obtained from sampling distributions, erroneous conclusions may occur. Given two alternative hypotheses, H_0 and H_1, there are two ways to be right and two ways to be wrong. (The customary names of these eventualities are given in Table 2.2.) The reader will encounter frequent references to the probability of making a Type I error, since this probability is set arbitrarily by the experimenter and in part determines the outcome of a statistical analysis. The Type I error probability is symbolized by the Greek letter alpha, α, or by an italic lower

Table 2.2 *Outcomes of statistical inferences*

		Actual situation	
		H_0 *true*	H_0 *false*
Decision	*Accept*	Correct acceptance	Type II error
	Reject	Type I error	Correct rejection

case p. It is also called the "level of significance" (see Sections 2.2.6 and 2.2.7) and typical preset values for α are .05 and .01. By turning back to Figure 2.1 (page 52) the reader can observe that p corresponds to one or both tails of the relevant sampling distribution. In the figure, .025 of the area is delineated in each tail, totaling to a p of .05 (.025 + .025) for the two-tailed test discussed in Section 2.2.6. If a one-tailed test were being made with only the upper tail of Figure 2.1, then p would equal .025. *By definition*, the p, or significance level, is equal to the probability that a sample mean will be of such magnitude as to lead to rejection of H_0 when it is in fact true. That is, p (or alpha) is the probability of a Type I error, which probability can be changed deliberately by moving the line of rejection in towards the mean (increasing p) or out from the mean (decreasing p).

The probabilities of the other three outcomes are determined in part by what is true in reality (the real value of μ) and in part by the preset p value. Note in Table 2.3 that for an actual situation ("in a state of nature"), statistical procedures allow only two decisions and that their probabilities sum to 1.00. Thus, if, when H_0 is true, it is nevertheless rejected a proportion, p or α, of the time, then it is correctly accepted $1 - \alpha$ $(1 - p)$ of the time.

The situation for a false H_0 is just a bit more complex. We can first note that the smaller α is, the farther from μ must $\bar{X}$ be to cause rejection of H_0. (See Figures 2.3a and 2.3b.) This means that if H_0 is in fact false, and μ is some value other than the hypothetical one, then it will be harder to detect this fact (that is, to correctly reject H_0) with smaller levels of significance. We see then that β, sometimes called "power," is smaller when alpha is smaller and that the probability, $1 - \beta$, of Type II error (failing to detect a real difference $\bar{X} - \mu$) becomes larger as α becomes smaller.

There is no simple inverse relation between α and $(1 - \beta)$. The relation is complicated by how large $|\bar{X} - \mu|$ (where μ is at its hypothetical value under H_0) *really* is relative to its standard deviation

Table 2.3 *Probabilities of statistical decision outcomes*

		Actual situation	
		H_0 *true*	H_0 *false*
	Accept	$1 - \alpha$	$1 - \beta$
Decision	*Reject*	α	β
		1.00	1.00

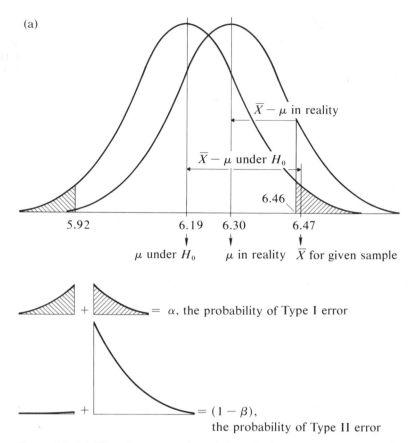

Figure 2.3 (a) *Visual presentation of the relation between probabilities of Type I and of Type II errors for a two-tailed test of a single mean hypothesis when N = 50 and σ is known (1.10). In this instance, no error occurred; the null hypothesis is correctly rejected.* (b) *Visual presentation of the relation between probabilities of Type I and of Type II errors for a two-tailed test of a single mean hypothesis when N = 30 and σ is known (1.10).*

$\sigma_X/\sqrt{N}$. Thus, if $(\bar{X} - \mu)$ is five times $\bar{X}$'s standard deviation, its $|z| = |\bar{X} - \mu|/\sigma_{\bar{X}}$ will be 5 $\sigma_{\bar{X}}/\sigma_{\bar{X}}$ or 5.00, and H_0 will be rejected almost no matter how small α is. That is, a z of 5.00 has a probability of occurrence under H_0 of 0.000 to three significant figures on the normal distributions. (A t of 5.00 has a probability of 0.000 also for $df \geq 7$.) On the other hand, if $|\bar{X} - \mu|$ were only one-half a standard deviation of the mean, we would have $Z = \frac{1}{2}\sigma_{\bar{X}}/\sigma_{\bar{X}} = .50$, and this value would lead to acceptance of H_0 at almost any common significance level,

(b)

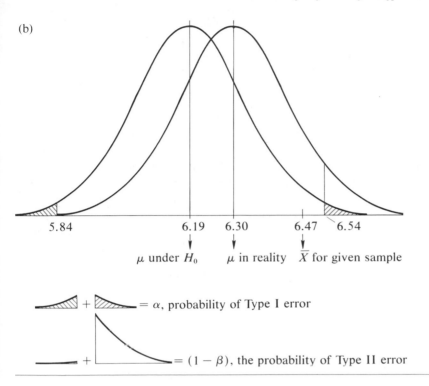

even though H_0 is literally false. Let us return to our example of Sections 2.2.6 and 2.2.7, in which H_0 calls for $\mu = 6.19$ and in which $\bar{X} = 6.47$. Suppose now that, contrary to fact, we could discover the real μ and, in so doing, we found it to be 6.30. Figure 2.3a shows the sampling distribution of means under H_0 (left curve) and in reality (right curve) for $N = 50$ while Figure 2.3b shows these same curves for $N = 30$. Note that for $N = 50$, a Type II error is not made, but that for $N = 30$ one is made. The bars on the H_0 curve determine where the rejection region—as it is called—will be. It will consist in the *base-line* values to the left and right, respectively, of these bars on the H_0 curve. Note that the reality curve actually determines the probability that a mean will or will not be in the rejection region. Thus, the appropriate area has been shaded *on the reality curve*, not the H_0 curve.

You are not going to see curves like this anywhere except in statistics books, and (under currently popular editorial policy) you will see only the slightest passing reference to Type II errors. However, a careful study of Figures 2.3a and 2.3b should impress upon you the joint roles in the final decision played by (1) the level of significance, (2) the sample size N, and (3) the magnitude of the difference between μ according to H_0 and the *real* μ.

In reading statistical reports, you may have occasion to ask, "How large would a difference have to be to be detected by the procedures employed?" Many basic texts give complete directions for finding this exact figure;[3] some readers will become adept at making an approximation to it. In any case, the question can be posed, "What is the smallest difference worth detecting?" The answer will vary greatly depending on the use to which the results are to be put. If acceptance of H_0 is going to halt all investigation in an area or if acceptance of H_0 is going to lead to the use of a potent new drug on the general population, slight effects should be detected. If H_0 is only a preliminary hypothesis, or if only relatively large[4] differences carry theoretical import, a higher probability of Type II error for small differences from H_0 can be tolerated. Again, it is a matter for your judgment—as well as the investigator's.

2.2.10 Hypothesis Testing: III. Differences between Means; Other Statistics

The test of the single mean is less common than tests of hypotheses regarding two means. Now that the logic of hypothesis tests has been outlined with reference to the single mean, we need only note that H_0 can be written in terms of the differences between two population means, μ_1 and μ_2, and that appropriate adjustments can be made in our six steps (pages 50–51).

1. Form H_0 so that the means do not differ; that is, $H_0: \mu_1 - \mu_2 = 0$ for a two-tailed test, and $H_0: \mu_1 - \mu_2 \leq 0$ for a one-tailed test. For a two-tailed test, the alternative is $H_1: \mu_1 - \mu_2 \neq 0$; for a one-tailed test, it is $\mu_1 - \mu_2 > 0$.

2. State the shape of the sampling distribution of the difference between means. This will be normal for a large N and t for a small N with normal distribution of the X scores within populations.[5]

---

[3] See, for example, W. Hayes, *Statistics for Psychologists*, New York, Holt, Rinehart, and Winston, 1963, pp. 204-206. In Hayes' discussion the difference is known, N unknown. In our example, N is known, the difference unknown.

[4] The notion of large difference having theoretical import and small difference not having such import depends on the existence of measurement adequate for the phrases "large enough" and "too small" to have some quantitative reference.

[5] See W. Hayes, *Statistics for Psychologists*, New York, Holt, Rinehart, and Winston, 1963, pp. 315-316.

3. Calculate the standard deviation of the sampling distribution of the difference between means. There are two distinctly different formulae for the *standard error of the difference between two means* (which is an estimate of the standard deviation of the sampling distribution just mentioned). One is for means from two independent groups—for example, experimental and control groups. The other is for two groups of correlated scores, as when there is one group and each subject has an experimental treatment score *and* a control score.

In the latter case, the two groups of scores are related, since a person may have some characteristic that influences both scores systematically. The investigator will usually not provide formulae and calculations (indeed, he need not!), but he should say which standard error term he is using. Phrases like a "*t*- (or *z*-) test for two independent means" or "a *t*-test (or *z*-test) for correlated means" are often used in the literature for this purpose.

4. Calculate the statistic to estimate $(\mu_1 - \mu_2)$, namely $(\bar{X}_1 - \bar{X}_2)$, where $\bar{X}_1$ is the mean of the experimental scores and $\bar{X}_2$ that of the control scores. Compare this with the hypothetical H_0 value of 0.

5. Calculate the ratio of

$$\frac{(\bar{X}_1 - \bar{X}_2) - (\sigma_{\bar{X}_1} - \sigma_{\bar{X}_2})}{\sigma_{\bar{X}_1 - \bar{X}_2}}$$

Where $\sigma_{\bar{X}_1 - \bar{X}_2}$ is the standard deviation (standard error) of the difference, $\bar{X}_1 - \bar{X}_2$. Under the null hypothesis, $\mu_1 - \mu_2$ is zero, and this ratio becomes

$$\frac{\bar{X}_1 - \bar{X}_2}{\sigma_{\bar{X}_1 - \bar{X}_2}}$$

6. Refer to the correct sampling distribution, normal or *t*; using a preset level of significance, decide whether $\bar{X}_1 - \bar{X}_2$ differs enough from zero to warrant rejection of H_0. Then, reject or accept H_0 as appropriate.

This general logic can be extended to the test of many hypotheses about population parameters that can be estimated by statistics whose formulae are known and that have sampling distributions with known shapes and standard deviations. Thus, tests are available for standard deviations, for differences between standard deviations, and for Pearson correlation coefficients (see Chapter 3), and for many other statistics. If the sampling distribution is neither normal nor *t* (it is neither for S or $\hat{S}$), then step 5 does *not* involve a ratio like that for means but another formula instead. All probabilistic logic is predicated on the correct choice of sampling distribution for some formula involving the statistic of interest. In later chapters, two distributions other than the normal and t will be considered.

2.3 Summary of Hypothesis Testing

The tests of hypotheses described in this chapter all follow the same logic: reject (decide to believe the falsehood of) that hypothesis which is improbable in light of the data. All of the tests require knowledge of the sampling distribution of the statistic (mean, difference of means) of interest, including its shape and its standard deviation (the standard error of the mean, of the difference between means, or whatever). This knowledge is gained from statistical theory and often requires several assumptions in addition to the hypothesis. If the statistic calculated to estimate the parameter is improbable in view of the assumptions, one of them—the null hypothesis—is rejected. The idea is that the statistic is here and based on observation. It cannot really be improbable; in fact, it is eminently probable. Therefore, something must be wrong with the series of steps by which we found it improbable. Since we've been careful with our other assumptions (always a statement open to question), let's reject the null hypothesis. In this testing process, it is possible to make two kinds of errors, deciding a true hypothesis is false (Type I) and deciding a false hypothesis is true (Type II). The probability of making a Type I error is set by the investigator and is called "the level of significance." The probability of making a Type II error is determined as a complex function of level of significance, sample size, and actual magnitude of the difference between the hypothesized value of the parameter and its actual value. For instances of fixed N and fixed real difference, the two error probabilities are related in an inverse manner. You should base your judgment of a test on whatever you can infer about the adequacy of the assumptions needed by the test used, and, more especially, on your own judgment of how different the actual parameter need be from the hypothetical H_0 value to be worth detecting.

You should also be alert to the fact that tests made on group data may not lead to conclusions universally applicable to all individual members of that group. The author's discussion should be carefully compared with the findings to see that this false impression is not created. Any inference about a single individual should be based on observation(s) of that individual. For example, the finding that a group of boys has an average aggression rating higher than that of a group of girls ($p < .05$) does not warrant such assertions as "Caleb will be more aggressive than Jennifer, since boys are more aggressive than girls." Nor are such statements as "A girl inhibits her aggressive impulses by sublimation," if "a girl" tends to imply *all* girls. Statements such as these may be helpful but they must be both used and interpreted with care. Similarly, any comparison of Caleb and Jennifer involves only the data for them. The group comparisons may properly be used to *suggest* an interpretation, if, in fact, Caleb is more aggressive than Jennifer. In turn, Caleb and Jennifer may be used to suggest interpretations regarding the group differences, but neither of these techniques should be offered as demonstrations of the certain validity of any conclusion.

References

Fawl, C. L., Disturbances Experienced by Children in their Natural Habitats. In R. G. Barker (Ed.), *The Stream of Behavior*. New York: Appleton-Century-Crofts, 1963, pp. 99–126.

Hayes, W., *Statistics for Psychologists*. New York: Holt, Rinehart, and Winston, 1963.

McClelland, D., and Apicella, F. S., A Functional Analysis of Verbal Reactions to Experimentally Induced Failure. *Journal of Abnormal and Social Psychology*, 1945 (40), 376–390.

3

Correlation for Readers

(1) Value of one measure reduces the uncertainty about the value of another measure

(2) Change in the value of one measure results in a change in the value in another measure

In this chapter we will consider some of the ways to demonstrate the presence or absence of a relationship between variables, and we will also consider when and how statements can be made about the extent of a relationship. To review, two sets of measures are related if knowledge of an individual's score on one of them reduces the range of possibilities for his position on the other one of them. Consider, for example, a measure of dependency that can range from 0 through 20, and a measure of compliance that can range from 0 to 15. Suppose that for individuals with a dependency score of 10, the compliance scores range only from 6 through 10. Only one-third of the available range of compliance scores is ever obtained by those with 10's on dependency. Thus, knowledge of the dependency position of an individual reduces the uncertainty about his compliance over what it otherwise would have been. This reduction in uncertainty can be phrased, "The dependency score of a person provides information about his compliance score." When two measures are so related, the term *correlation* is usually used to describe the fact. There are various ways to demonstrate the presence or absence of correlation between two measures. Some of the techniques also permit their users to make assertions about the degree or magnitude of the relationship between the measures.

3.1 Detecting Correlational Relationships in Use

It is important for a reader to keep in mind the distinction between (*a*) presence or absence of relationship and (*b*) degree of relationship. Most readers will find that their qualitative interpretations and their conclusions from reading correlational studies depends on their personal skill in interpreting results. Let us begin with the techniques for finding out whether two measures are or are not correlated.

3.1.1 Group Differences

In many studies in which pre-existing groups are compared to each other, the feature that distinguishes one group from another defines one of two variables entering into a correlational relationship. The measurement on which the groups are compared is the other variable. Suppose, for example, that we observe Table 3.1 in an article. Notice (in the first row) that the mean dependency score for women is higher than that for men. Suppose further that the author has reported that the mean difference is such that it would occur by chance less often than once in a hundred times.

We may then conclude that sex and dependency—at least as defined by this score—are correlated. This means a relationship exists, but we cannot be sure how great it is. We can say that knowledge of an individual's sex reduces the range of possibilities for his dependency score. Given the second row of the table, we can even make an educated guess about how much reduction of uncertainty has occurred. Usually the range is covered by five standard deviations

Table 3.1 *Mean dependency scores for students in Psychology 300*

Dependency	Sex	
	Male	*Female*
Mean	9.12	13.28
SD	1.04	2.02
N	157	89

(2.5 S above the mean and 2.5 S below; see further, Chapter 1). We thus predict the men's effective range to be, in whole numbers, from 6 to 12 ($9.12 \pm 2.5 \times 1.04$) and the women's from 10 to 16 ($13.28 \pm 2.5 \times 2.02$). Thus, it would be extremely rare for a woman to score 8 and not so surprising for a man. Conversely, it would be rare for a man to score 13, which is approximately the women's average.

What is established in this table is that a relationship exists between a certain dependency score and sex. Even though a *critical ratio* or value of *t* may be presented, it is the establishment of a correlation that is being made.

Sometimes, there is more than one group, as when groups are defined by age; this means the interpretation may be more complex. Let's look at Table 3.2. It appears that the dependency scores of boys increase systematically with age. If so we can say that a correlation exists between age and this measure of dependency. We can further assert that the correlation was positive, which means that higher ages go with higher scores and lower ages with lower scores.

This assertion—that dependency increases with age—might seem strange to, say, a student of personality development. He might not know what to make of it. Perhaps he should look more closely at the dependency score. It may really be a measure of need for peer acceptance, a characteristic which a student of development would expect to see increase during childhood.

How would the reader find this out? The author has provided him with evidence that *a* correlation exists between two variables, age and dependency. One of these variables is defined (measured) in years of age at last birthday. The other variable is measured by some kind of psychological test score that is defined by items and the method of scoring responses. By reading further in the text of the article, the reader may discover that items such as "likes to be with someone else while working" and "seeks adult approval" are typical. Perhaps teachers or other supervising adults mark these items for each child. These facts would suggest to a student of personality development that it is "need for being with others" that is measured.

Table 3.2   *Mean dependency scores for boys. First example*

Dependency	Age at last birthday					
	2	*4*	*6*	*8*	*10*	*12*
Mean	8.1	8.58	9.10	10.76	11.51	12.38
N	20	20	20	20	20	20

Table 3.3 *Mean dependency scores for boys. Second example*

Dependency	Age at last birthday					
	2	4	6	8	10	12
Mean	8.1	12.38	11.51	9.10	10.76	8.58

In other articles, no examples of test items may be given, and the reader can only speculate on what may be the cause of the presence of correlation between two variables that do not seem to go together. Personal judgment here is unavoidable—in fact, it is desirable. Logic must be applied to the described group differences to provide a meaningful interpretation of numerical results.

It is also possible in the case of multiple groups for the differences between pairs of groups to form no meaningful pattern. For example, suppose Table 3.2 looked, instead, like Table 3.3. In this study, the two-year-olds are *less* dependent than the four-year-olds, who are *more* so than the eight-year-olds. There does not seem to be any logic to these differences. That is, it appears as if the boys are getting scores that are not far different from random drawings from a hat containing slips of paper numbered 1 through 20. If this is really so, we would say the dependency measure is unreliable. This statement is suggested, but *not proved*, by the data of Table 3.3. Clearly, the reliability of the measures should be examined.

3.1.2 The Contingency Table

A frequently occurring type of correlational study is one in which each of two variables is defined by group membership. Consider again the possible relationship between age and dependency. We define age as before, providing four groups. Instead of our hypothetical dependency test of Tables 3.1, 3.2, and 3.3, let us suppose that we had the boys' teachers categorize them as being above average, average, or below average in dependency. Then the dependency variable provides three groups.

Membership in any one of the four age groups could, in principle, be combined with membership in any one of the three dependency groups. This provides a total of twelve (4 × 3) groups into which N boys could be classified. The relationship (or lack of it) between two variables defined by groups is shown in a table, such as Table 3.4, called a contingency table. Table 3.4 has three rows representing dependency and four columns representing age. The numbers in the

Table 3.4 *Classification by age and teacher's dependency ratings*

Dependency	Age				Row totals
	2	*4*	*6*	*8*	
Above average	15	10	5	0	30
Average	10	10	15	5	40
Below	0	5	5	20	30
Column totals	25	25	25	25	100*

*Grand total.

cells are frequencies and represent the numbers of boys who are of the column age and row dependency scores.

The chi-square statistic, χ^2, is often computed to establish the presence or absence of a relationship between the row classification and the column classification in a contingency table. The size of a chi-square statistic varies directly with the extent to which the pattern of frequencies in the columns differ from the over-all pattern for the row totals.

If the proportion of below-average, average, and above-average dependent boys is nearly the same for two-year-olds, six-year-olds, and the total sample of 100 boys, then chi square will be small, and no relationship will be said to exist. If, however, there are disproportionately (to the row totals) more above-average two-year-olds, average four-year-olds, and below-average six-year-olds, then chi square will be large, and it can be concluded that a relationship exists between dependency and age. For Table 3.4, the value of the chi-square statistic is 38.33; since a value at least that large occurs by chance less than one time in one thousand, we conclude that a relationship exists.

To make this clearer, let us rewrite Table 3.4 with proportions of the column totals in the cells instead of frequencies. These figures appear in Table 3.5.

Table 3.5 *Column proportions corresponding to Table 3.4 frequencies*

Dependency	Age				Over-all pattern
	2	*4*	*6*	*8*	
Above average	.60	.40	.20	.00	.30
Average	.40	.40	.60	.20	.40
Below average	.00	.20	.20	.80	.30
Column totals	1.00	1.00	1.00	1.00	

Clearly, as we go from left to right (younger to older), the proportion of below-average boys increases, while the proportion of above-average boys decreases. The proportion of average boys increases and then decreases. These patterns tell us that older boys are judged less dependent by their teachers than younger boys. This is reflected in a high chi-square value.[1]

A chi-square value large enough to attain some preselected level of significance permits only the conclusion that the two variables are related. No statement can be made from a χ^2 statistic alone about the extent of a relationship. The χ^2 values from two different contingency tables should *never* be used for the purpose of comparing the extent of relationship between variables A and B with the extent of relationship between C and D. Such inappropriate comparisons are sometimes encountered, and you are warned against accepting them.[2]

3.1.3 Scatter Diagram

If each of the variables under study is measured on a scale with many possible values, then a contingency table representing their relationship would have a large number of cells. To avoid constructing such a complex table, a *scatter diagram* is often constructed instead. Such a diagram has vertical and horizontal axes like a graph. The vertical axis represents one of the variables and is marked off into the units used to measure that variable. The horizontal axis is similarly marked off into units representing the other of the two variables. Figure 3.1 is a scatter diagram that shows the relationship between age in years and dependency, as measured by a test like the one of Tables 3.1, 3.2, and 3.3.

Each cell of the diagram is defined by a column number (an age) and by a row number (a dependency score). Thus a given tally repre-

---

[1] The definition formula for χ^2 in terms of frequencies makes it clear that large deviations of frequencies from those expected on the basis of over-all patterning contribute large terms to χ^2 whereas small deviations contribute little to it. Here is the formula:

$$\chi^2 = \frac{\sum (0 - E)^2}{E}$$

where summation is over all cells and $0 =$ the cell frequency such as 15 in the upper left cell of Table 3.4 and $E =$ the cell frequency expected for that cell if the patterns on the right "total" column were followed. The expected value for the upper left hand cell of Table 3.4 would be 30/100 times the total of 2-year-olds $= 30 \times 25 = 7.50$.

[2] There is a statistic called the "contingency coefficient," derived from chi square, that can, under certain assumptions, be used to describe the degree of relationship in a contingency table. It is discussed in Chapter 6.

sents the joint occurrence of a given age and a given dependency score. Using this idea, we see that the diagram is really a joint frequency distribution. We could, if we liked, list pairs of ages and dependency scores and place their frequencies next to them. This would produce a column such as that for a distribution of scores defined only by age or only by dependency score. The column would be long and the procedure would be cumbersome because of the many pairs of scores. The reader should note its *logical* similarity to a scatter diagram, however.

Now, of what use is a scatter diagram? By inspecting it we can tell if there is a tendency for tallies to cluster in a specifiable fashion. Three patterns would be of special interest.

1. If the tallies are concentrated in the lower-left corner, center, and upper-right corner of the diagram (Figure 3.1), we could conclude that there is a tendency for low dependency to go with low age,

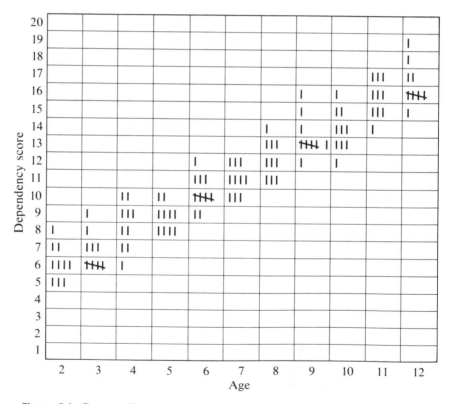

Figure 3.1 *Scatter diagram relating dependency score and age for 90 boys.*

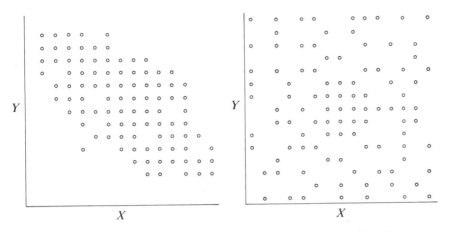

Figure 3.2 (*a*) *Scatter diagram for a negative* (*inverse*) *relationship between X and Y.* (*b*) *Scatter diagram showing no relationship between X and Y.*

moderate dependency with moderate age, and high dependency with high age. This could be restated, "There appears to be a direct [positive] relationship between age and dependency." This conclusion would correspond to the impression one gets from Table 3.2.

2. Another pattern of special interest would be that in which the scores are concentrated in the upper-left, center, and lower-right portions of the diagram (Figure 3.2a). This pattern would indicate that low age goes with high dependency, moderate age goes with moderate dependency, and high age goes with low dependency. This is stated, "There appears to be an inverse [negative] relationship between age and dependency."

3. A third pattern of interest would be one in which the tallies are not concentrated anywhere in particular, and some appear in all portions of the diagram (Figure 3.2b). We would get the (usually correct) impression from this pattern that there is no relationship between age and dependency.

3.2 Pearson *r*

The Pearson product-moment correlation coefficient, r, is often used with data appropriate for scatter-diagram presentation. It has a known distribution and thus the hypothesis of no relationship can be tested according to the logic of Chapter 2. Under appropriate circumstances (described below), r can be used to describe the degree of relationship between two variables of the kind for which a scatter diagram could be used.

3.2.1 Definition Formula

Pearson r is defined in this manner:

$$r = \frac{(X - \bar{X})(\bar{Y} - \bar{Y})}{NS_X S_Y} \tag{3-1}$$

where

$\bar{X}$ = Mean of the X variable
$\bar{Y}$ = Mean of the Y variable
S_X = Standard deviation for the X variable
S_Y = Standard deviation for the Y variable
N = Total number of pairs of scores

So defined, r varies between -1.00 and $+1.00$.

By convention, X is usually represented on the horizontal axis of a scatter diagram; Y on the vertical. For correlational work, the distinction is arbitrary and does not imply that Y depends on X in any causal way. Equation 3-1 shows us several things: (1) If $X - \bar{X}$ and $\bar{Y} - \bar{Y}$ both have "$+$" signs or both have "$-$" signs, then the product will be positive. If this tendency holds for most pairs, r will be high (closer to the $+1.00$ end of its scale). This corresponds to Figure 3.1. (2) If, instead, most pairs of scores produce $(X - \bar{X})$ and $(Y - \bar{Y})$ of opposite signs, the products will be negative, and r will be near to its -1.00 scale point. This corresponds to the diagram in Figure 3.2a. (3) If $(X - \bar{X}_X)$ and $(Y - \bar{Y})$ have the same sign about as often as they have opposite signs, the sum will be low positive or low negative, and r will be near to 0.00. This is true in Figure 3.2b. Thus, the three situations presented in Figures 3.1, 3.2a, and 3.2b have their algebraic counterparts in the formula for r. So do moderate amounts of correlation between -1.00 and 0; and between 0 and $+1.00$.

3.2.2 Interpretations of r

The reader of the literature, when confronted with an r or r's, presumably wants to know what to make of it. There are several ways in which r can be used for this purpose; each requires that certain assumptions be met. The reader of a contemporary journal will often have to rely on incomplete information in deciding whether a given interpretation of r is appropriate.

3.2.2.1 Variance. The most useful reader interpretation of r relies on the fact that r^2 is equivalent to the proportion of variance (S^2) in one of the variables which is shared by the other.[3] Thus an r of .30 between age and dependency would suggest that they share $(.30)^2$ or 9 percent $(100 \times .09)$ of their variance—and that 91 $(100 - 9)$ percent of the variance in dependency is due to variables other than age. The idea is that any variable, Y, can be in a sense defined by one or more other variables, X. The square of r_{xy} indicates proportionately how great a role X plays in defining Y.

This use of r^2 requires (in its derivation) the assumption that the clustering of tallies on the scatter diagram can best be described by a straight line rather than some form of curve. Unless a scatter diagram or the lists of scores are shown, it is difficult to verify this assumption directly. It should be noted that r^2 will *underestimate* the shared variance if the assumption is violated.

3.2.2.2 Significance. The author of an article will often test his r's for significance. Almost any r can be shown to differ from 0 if N is large enough.[4] Thus the reader will need to decide for himself if an r of given magnitude is significant in any sense besides statistical. It is suggested that r^2 be considered in making this decision. If a noticeable proportion of variance in one variable, X, can be explained by the other, Y, then indeed Y helps the reader understand the meaning of X, and vice versa. If the proportion of shared variance is low, then no such conceptual help is rendered.

3.2.2.3 Prediction. In many instances it is the investigator's purpose to predict the score for an individual on one of two correlated variables from knowledge of the other. For example, he may wish to predict an experimental subject's score on an achievement test (Y) from knowledge of his intelligence as measured by a standard test (X). Or he may be interested in achievement motivation as a factor in test-taking and thus wish to predict X from Y.

If the relationship between X and Y is best described by a straight line, the algebraic equation of that line,

$$Y' = B_Y X + A_Y \tag{3-2}$$

may be used for the desired prediction. Such an equation is called

---

[3] See McNemar, Q., *Psychological Statistics*, New York, Wiley, 1965, pp. 129 ff.
[4] This is because the standard error of r depends only on N. For $N \geq 50$ it is $1/\sqrt{N}$ providing H_0, $r = 0$, is true.

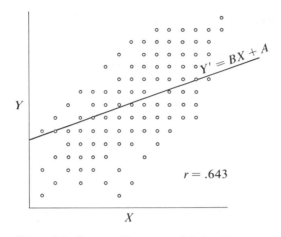

Figure 3.3 *Scatter diagram with its "best fit" regression line.*

a "regression equation". In a regression equation, the constants can be shown to be

$$B_Y = r\frac{S_Y}{S_X} \tag{3-3}$$

$$A_Y = \bar{Y} - B_Y\bar{X} \tag{3-4}$$

where, as usual, S_X is one sample standard deviation for X and S_Y is the corresponding standard deviation for Y. $\bar{X}$ and $\bar{Y}$ are the sample means for the two variables. An example of a scatter diagram with its corresponding equation is shown in Figure 3.3.

3.2.2.4 Error of prediction. Y has a prime sign (') by it in Equation 3-2 to indicate that the value of Y obtained *is not exactly equal to the individual's actual Y score.* Instead it is the best prediction we can make about Y, given (1) knowledge of X, and (2) a linear relationship between X and Y.[5] Unless $r = 1.00$, there will be some discrepancy between Y' and Y. This is the crux of the matter.

We have already seen in Chapter 1 that without any knowledge of X, our best[6] bet for Y is $\bar{Y}$. Further, we discovered that if we bet on $\bar{Y}$ for all persons, the best[7] estimate of our error was S_Y, the standard deviation of Y. In order for X to be of much help in predict-

---

[5] Technically, use of X to predict Y (or vice versa) also requires that we accept a particular definition of "best" called the "least-squares criterion." This concept will not be further discussed here.

[6] See note 5.

[7] See note 5.

ing Y, it should predict Y' so that Y' deviates from Y, on the average, less than $\bar{Y}$. Otherwise, we might as well forget about X and just use $\bar{Y}$.

Now, how shall we tell how much Y' betters $\bar{Y}$ as a prediction of Y? For any given case, we would just have to look at Y and check. But if we are willing to assume that the variances of Y about Y' are equal for all values of X, then we can determine an estimate of the error involved in using X to predict Y via the regression equation, Equation 3-2. This error is given as

$$S_{Y-Y'} = S_Y \sqrt{1 - r^2} \qquad (3\text{-}5)$$

The main thing you should note here is that we have a reduction in error (over that incurred by betting $\bar{Y}$ for all cases) estimated by the value $\sqrt{1 - r^2}$. The numerical value of r must be quite large before this reduction is noticeable. For example, to obtain a 10 percent reduction in error, r must be .44 or greater (to the nearest hundredth). This value was obtained by equating the value .90 $(1.00 - .10)$ and $\sqrt{1 - r^2}$ and solving for r.[8] A 25 percent reduction requires r to be at least .67, and a 50 percent reduction requires r to be at least .87. Thus, if you are adept at mental arithmetic (a handy trait for careful readers of the psychological literature), you can make a rough estimate of how much error reduction is being gained by the use of a regression equation in the prediction of an individual's score on one variable from his score on another variable.

In general, the use of regression equations in making decisions about individuals is not warranted unless r is quite high. This is especially true if the decisions are important to the individual's personal life, as would be getting a job, being admitted to college, obtaining specialized training in the armed forces, and so forth. As always, the question "How high?" becomes one of judgment. The reader will have to evaluate conclusions based on regression equations on the basis of his judgment about "How high is high?" in the given situation.

3.2.3 Reader Evaluation of the Adequacy with which Assumptions are Met

At this point you may well be saying to yourself, "I don't see many scatter diagrams in journal articles and books, only in statistics textbooks. How am I supposed to know if the journal and book authors know what they are doing?" There is no sure way. The

---

[8].90 is $1 - .10$ and would represent a 10 percent reduction.

conservative solution is to note that conclusions based on Pearson r's are rarely *more* adequate than ones truly justified by the mathematical theory on which r is based. For example, the principal assumption made in using r is that of linearity of relationship. Note that even if a line is a poor fit to a scatter diagram, there is always *some* line that is the best fit.[9] The use of this line for interpretive purposes will produce high error of prediction, which the reader can spot by mentally approximating[10] Equation 3-5. Further, such inappropriate lines are associated with low values of r.

The reader's main task is to spot high r's and to be skeptical of overinterpretation of low r's. I will have to leave him on his own with the guidelines of Section 3.2.2 to decide "How high is high?"

3.2.4 Reasons for Small r's

When we encounter a low r, the most immediate inference is that the two variables are unrelated (or only slightly related). This conclusion is usually correct. However, we shall now examine some situations in which a calculated r can be low even when there actually exists a relationship between the two variables. The reader will want to use this information whenever he encounters an r that is unexpectedly low; that is, when theory or previous findings have led him to anticipate a moderate to substantial r value.

3.2.4.1 Variance of a third variable. The first of these situations is one in which each of the two variables (X and Y) contributing to the value of r are related to a third variable, V, *but in opposite directions.* That is, r_{VY} might be negative and r_{VX} positive. If V has been allowed to vary over its entire range, r may have been depressed. This situation is presented in Figure 3.4 in a somewhat exaggerated way to show the point. The correlation between X and Y for the entire group is negative, and the line for predicting Y from X for the entire group of 80 subjects (not shown) would go from the upper-left portion of the figure to the lower-right. The three lines shown represent the corresponding regression lines for subgroups formed on the basis of V scores. When the value of V is held relatively constant within subgroups defined by the V score, then the expected positive relationship between X and Y appears for all three groups. The exact values of relevant r's appear next to the regression lines in Figure 3.4.

---

[9] See note 5.

[10] For such mental approximation, it is well to remember that the square root of a decimal value less than 1.00 will be greater than that value.

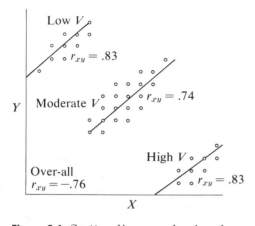

Figure 3.4 *Scatter diagram showing the relationship between* X *and* Y *when each relates to* V, *but in* opposite *directions.*

3.2.4.2 Moderator variable. The situation shown in Figure 3.5 is related to that shown in Figure 3.4, but the two situations are not identical. It may be that the correlation postulated between X and Y holds only for some of the people in the group. We can start with the assumption that it holds best for those individuals whose actual Y scores are close to the Y scores predicted for them from their X scores. In a visual picture of the situation, such as Figure 3.5, these are the individuals whose tally marks lie close to the regression line.

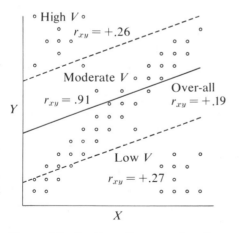

Figure 3.5 *Scatter diagram for the "moderator-variable problem."*

Arbitrarily, lines have been drawn parallel to the regression line, one at a distance of $S_{Y-Y'}$ above it, and one at the same distance below it. The individuals within these confines are called "predictables."

Now, suppose the "unpredictables" can be shown to differ systematically from each other and from the predictables according to their respective scores on some third variable, V. We could then think of V as a predictor of predictability.

For example, suppose X were an aptitude test, Y were a job-success score, and V were a measure of need to achieve. Figure 3.5 depicts a situation in which persons of low aptitude (X) *but* high achievement motivation (V) get high (that is, higher than predicted) job-success scores (Y) and, conversely, those with high aptitudes (X) and low achievement motivation (V) get low (that is, lo*wer* than predicted) job-success scores (Y). Those of moderate achievement have job-success scores (Y) that accord well with those predicted by aptitude (X).

Achievement motivation (V) appears to moderate the effect of aptitude (X) on success (Y); hence the term "moderator variable" for it. For any such V, if we divide the total group into subgroups on V, we will invariably find r_{XY} higher in one or more of the subgroups than for the group as a whole. In Figure 3.5, those above the top broken line have $r_{XY} = .26$, those within the broken lines have $r_{XY} = .91$, and those below the bottom broken lines have $r_{XY} = .27$. The over-all $r_{XY} = .19$.

This problem differs from the first one (Section 3.2.4.1) in that no conditions are put on the relationships between V and X and between V and Y.

If you encounter an unusually low r, you will want to refer to the author's description of how the sample group was composed. Perhaps there will be a hint as to what might be functioning as a moderator variable. Usually, scores on this variable will not be available and the identification of it cannot be verified by the reader. It would simply serve as a hunch to use in trying to synthesize the article at hand with several sources, all of which report moderate-to-high correlations between X and Y. If further study is to be planned, the scores on V would have to be found, then subdividing done, the several r's computed, and the hunch verified or abandoned.

3.2.4.3 One variable artificially restricted. In the third situation, in which r_{XY} may appear to be smaller than it really is, one of the variables is artificially restricted. This is shown in the scatter diagram of Figure 3.6, in which X stands for scholastic aptitude, and Y stands for college grades. If X is used in selecting freshmen for college A, then the population of freshmen at college A will have a considerably smaller range of scores on X than will the population of all individuals who take the scholastic-aptitude test (X). A partition has been drawn in Figure 3.6 at the point college A cuts off admissions. Notice that the line that would best fit the configuration of tallies to the right of the partition is much more nearly horizontal than the regression line for the entire group. In this example, this means that there will be no (low) correlation between X and Y whenever X has been used for selection. In all situations in which X is used for selecting people, a reduction in r_{XY} for those selected will occur. The magnitude of the reduction varies directly with the magnitudes of the differences between S_X^2 for the original group and S_X^2 for selectees and between S_Y^2 for one original group and the selectees.

In other instances, the range of X may have been restricted in ways that are not so obvious as the selection of candidates for a program of training or for employment. There may have been sampling procedures employed that had the effect of restricting the range on X, as when X is height and college basketball players are used as subjects. Your job in integrating several studies, only one of which reports a low r_{XY}, is to look closely at the subjects. You must ask, "Are they from the same populations, at least as regards the range of X?" If not, consider the possibility that the range of X may be restricted in the one study. Similarly, depressing effects on r may result from restriction on the range of Y or of the ranges of *both* X and Y.

3.2.4.4 Nonlinear relationship between X and Y. The calculation

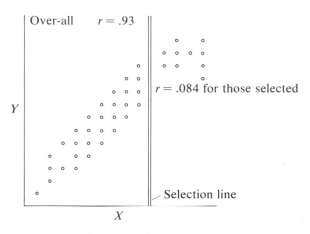

Figure 3.6 *Scatter diagram for the "restriction-of-range problem."*

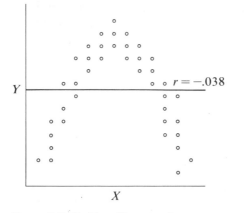

Figure 3.7 *Scatter diagram for a non-linear relationship between X and Y.*

of r is designed to produce a value that best[11] describes the extent of *linear* relationship between variables X and Y. The formula can be followed, even if the variables are related in a curvilinear fashion, as shown in Figure 3.7. In such a case, r will be quite low even though there is a systematic correspondence between X and Y.[12]

You might suspect that this explanation is correct if studies other than the one you have at hand have reported a curvilinear relationship—or if theory predicts that form of relationship.

3.2.4.5 Unreliability. A last, and technical, reason for r_{XY} to be depressed is that one or both measures are unreliably measured. This happens frequently, and thus deserves mention. It will not be further discussed.[13]

3.2.5 High values of r.

The first reason for obtaining an r in the .80's or .90's would, of course, be that there exists a strong positive linear relationship between X and Y. Here we will discuss two situations in which high values of r are calculated and reported, even though no meaningful relation exists between X and Y.

3.2.5.1 Item overlap. The first of these is a situation in which two correlated test scores, X and Y, utilize many of the same items. If many of the items in the test for X appear also in the test for Y, a high correlation will necessarily result, but will hardly be in-

[11] See note 5.
[12] The relation is best fit by the equation of a parabola.
[13] See McNemar, Q., *Psychological Statistics*, New York, Wiley, 1965, pp. 153 ff.

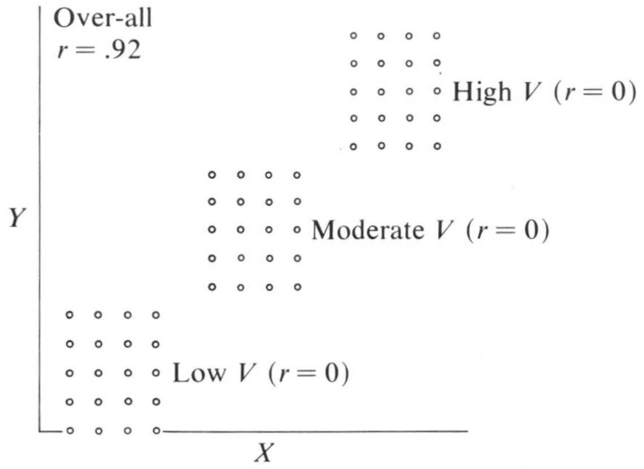

Over-all
r = .92

○ ○ ○ ○
○ ○ ○ ○
○ ○ ○ ○ High *V* (*r* = 0)
○ ○ ○ ○
○ ○ ○ ○

Y

    ○ ○ ○ ○
    ○ ○ ○ ○
    ○ ○ ○ ○ Moderate *V* (*r* = 0)
    ○ ○ ○ ○
    ○ ○ ○ ○

○ ○ ○ ○
○ ○ ○ ○
○ ○ ○ ○ Low *V* (*r* = 0)
○ ○ ○ ○
○ ○ ○ ○

X

Figure 3.8 *Scatter diagram showing the relationship between* X *and* Y *when each relates to* V *in the* same *direction.*

formative. For example, if *X* consists of pass-fail items 1, 2, 3, and 4, and *Y* of items 1, 2, 3, and 5, then 75 percent of the *X* score (total passes) is exactly equal to 75 percent of the *Y* score. A low value of *r* would not be possible in such an instance nor would the high value really be informative. You should conclude that two measures of the same variable are being shown.

3.2.5.2 Influence of a third variable. The second situation in which *r* may be misleadingly high is the one in which *X* and *Y* both increase or decrease with a third variable, *V*, as is shown in Figure 3.8. In this exaggerated presentation, *X* and *Y* superficially seem to correlate. But, on closer inspection, we see that there are three separate concentrations of tallies; that within each concentration, *X* and *Y* *do not* correlate. This means that if the range of *V* were narrowed, that is, held relatively constant, then *X* and *Y* would not appear to correlate at all; the apparent correlation we initially observed was due to their having *V* in common. When the reader encounters an r_{XY} that he needs to understand, he should try to propose a *V* variable that *X* and *Y* might have in common. If he is correct, then when *V* is held constant in a group, r_{XY} will disappear; that is, under these circumstances, r_{XY} will be small. So, if one study finds high r_{XY} and another study finds r_{XY} near 0, one of the possible reasons could be that *V* varies a lot for the high r_{XY} sample and is relatively constant in the other. Such an interpretation requires that the variation of *V* in the samples reflects the variation of *V* in the reference populations.

3.2.6 The Problem of a Small Number of Possible Scores

The derivation of r assumes continuous scales for X and for Y. This means that relatively many values of X and Y are anticipated. In practice, r is sometimes applied to X and Y variables measured on scales with very small numbers of possible values. This type of application distorts r in the sense that it indicates a different degree of relationship than would be found if a greater number of score values were possible. Whether $r_{pop.}$, the population value corresponding to r, is underestimated or overestimated cannot be predicted with precision from information about the number of scale points. We can get an idea of what happens by taking a look at an extreme case in which X and Y are both measured on two-point scales. There are only four possible pairs of scores, 0,0; 0,1; 1,0; and 1,1. The possible values for r are severely limited relative to the range of decimal values between -1.00 and $+1.00$—all of which are usually candidates as possible r values. If all N subjects received the same pair of scores, then r, according to the usual formula, would be "undefined" $(0 \div 0)$. If the pairs of scores distributed themselves equally among the four possibilities, as in Figure 3.9, r would be 0.00, as expected. If the scores distributed themselves between the upper-right and lower-left points equally, r would be 1.00, as expected. For equal distribution between upper-left and lower-right points, r would be -1.00, again as expected. But for all intermediate situations, r would not yield an accurate picture. If all pairs of scores except one were at the upper-right point and that one were at the upper-left point, r would again be undefined.

If all but two were at the upper right, one at the upper left, and one at the lower right, r would be equal to $-1/(N-1)$, which approaches 0.00 as N becomes large. If all but two of the scores were divided between the upper right and the lower left, with one at the

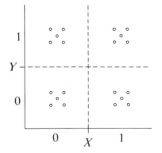

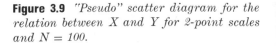

Figure 3.9 *"Pseudo" scatter diagram for the relation between X and Y for 2-point scales and $N = 100$.*

upper left and one at the lower right, r would be calculated as $1 - 4/N$, and if N were 4, r would be 0.00. In any case, the calculated value of r is apt to be misleading.

Now, before we wring our hands in despair, let us observe that there are a variety of other, more appropriate, methods of calculating the extent of correlation for variables with few score values possible. You may not want to be burdened with all of their technical details, but you do need to know that they exist and to know their names, so that you can better understand authors who use them—or should use them.

If the score values represent broad categories into which continuously measured scores are classified, the *tetrachoric correlation coefficient*[14] is appropriate. If this is true for only one of the variables, the other being measured continuously, the *biserial correlation coefficient*[15] is appropriate. If both variables are truly two-point discrete measures, the *phi coefficient*[16] is appropriate; if one variable is a true two-point discrete measure, and the other a continuously measured variable, the *point biserial coefficient*[17] is appropriate.

3.2.7 Presentation of Many r's from the Same Sample

In field studies (see Chapter 2), one often sees a large number of characteristics measured for the same group of subjects. These measures are then correlated with each other, resulting in a large number of r's. Discussions then follow in which the pattern of r's is interpreted by the author. Usually some size is selected which r must attain before it is interpreted as part of a pattern (minimum size for statistical significance or some other size considered meaningful). Note that this procedure always dictates—via statistical theory—a certain probability, p, that any given r will exceed the criterion size even though there is no relationship between the variables. When the number of r's is large, it is well for the reader to remember that $(100\,p)$ percent of the r's interpreted could attain the size required for significance by chance. There is no way to pick out *which r's are the culprits*. One must avoid picking as culprits those r's that do not fit in with some explanation of the data that would otherwise be feasible. The chance of error can, of course, be reduced

---

[14] See McNemar, Q., *Psychological Statistics*, New York, Wiley, 1965, Ch. 12.
[15] *Ibid.*
[16] *Ibid.*
[17] *Ibid.*

by requiring high values of r before interpretation. As always, the intelligent interpretation of the meaningfulness of such data depends on the reader's judgment.

3.3 Regression to the Mean

Frequently, articles report that the means for a second measurement of some group are lower than the first means were; this is likely to occur in any study where there are pretests and posttests. You may see this phenomenon described as an instance of "regression to the mean." This phrase describes a process that can be explained in the following way:

Suppose a linear relationship exists between two variables, X and Y, but that this relationship is not perfect; that is, r is neither plus or minus 1.00. It is a mathematical fact that a Y score predicted from an X score will necessarily be nearer, in standard-deviation units, to $\bar{Y}$ than X is to $\bar{X}$. This can be seen by examining a correlation scatter diagram and the best-fit line for prediction for standard scores. In Figure 3.10, such a diagram is presented for $X =$ first-test IQ score and $Y =$ second-test IQ score. In this figure, the variables

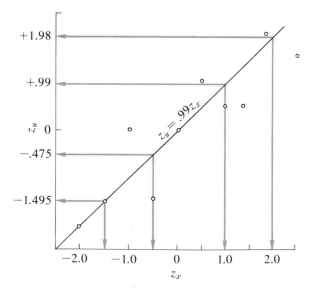

Figure 3.10 *Correlation diagram for two standard scores in which $r = .99$.*

that actually appear on the two axes are standard scores, z_X and z_Y, standing for the standardized versions of X and Y, respectively. Several specific z_X values have been identified, and their corresponding predicted z_Y's indicated. Notice the equation of the line $z'_Y = r_{XY} z_X$.[18] It is immediately apparent that the predicted z'_Y must be smaller in absolute size than z_X, unless r_{XY} is 1.00 or -1.00. This entire phenomenon is termed "regression to the mean."

Historically, the term "regression" was used in discussing all problems of correlation and prediction, primarily because of the phenomenon we have just mentioned. Indeed, r stands for "regression." Early students of behavior thought that successive generations of organisms, when measured on the same trait, were becoming more and more mediocre and that individual differences were being eliminated due to regression effects. This is obviously not happening, because, if it were, we would all by now have approximately the same IQ's, heights, pulse rates, and so forth.

The key to the illusion is this: Actual z_Y scores do not conform perfectly to predicted z'_Y scores, just as actual Y scores do not conform perfectly to predicted Y scores. So, for all persons with a stated z_X score, the actual z'_Y scores will range from the predicted z_Y to values both larger and smaller than it. For large predicted absolute values of z_Y, there will necessarily be more actual z_Y scores smaller than their corresponding z_X scores in absolute value than there will be actual z_Y scores larger. This is because for very large z's, there is nowhere to go but down. This explanation is illustrated in Figure 3.11.

On the horizontal axis appear values of z_X presumed to represent standard scores for some kind of pretest. The vertical axis represents the frequency for large N. Thus, the curve shown is an idealized frequency-distribution curve of the normal-curve shape. Within each interval, there are miniature curves representing the hypothetical distributions of *actual* z_Y = posttest scores. Notice that these distributions seem at first glance to be skewed for the extreme intervals. However, we must also observe that some of the pretest scores nearest the mean are contributing posttest scores farther from the mean. A process of compensation is going on that results in a z_Y

---

[18]This equation can be derived from Equations 4-3, 4-4, and 4-5. First substitute z_X for X and zy for Y in formula 4-5, obtaining

$$z'_Y = B_{z_Y} z_X + A_{z_Y}$$

And then, from substitution of $S_{z_Y} = S_{z_X} = 1$ and $\bar{z}_Y = \bar{z}_X = 0$ into Equations 4-3 and 4-4 for B_{z_Y} and A_{z_Y}, respectively, we have $z'_Y = r_{XY} z_X$.

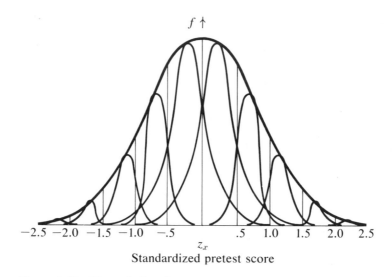

$f \uparrow$

-2.5 -2.0 -1.5 -1.0 -.5   .5   1.0   1.5   2.0   2.5

z_x

Standardized pretest score

Figure 3.11 *Normal distribution of pretest scores showing the distribution of posttest scores within each pretest interval.*

distribution (not shown for the entire group) identical with the one for z_X. If the relationship is not perfect, as it is not in this example, then this compensation will not be perfect. The mean of the z_Y scores for any of the extreme intervals of z_X will indeed be nearer to 0 than was the z_X value marking the midpoint of the interval.

The reader is likely to encounter references to "regression effects" in discussions comparing two extreme-scoring pretested groups with respect to posttest scores for the same variable. The classic example is this: Suppose we administer an IQ test to a population of preschool children. We then provide enrichment training (nursery school, Head Start, or whatever) to the lowest-scoring children and provide no special training to the average or high-scoring children. We then administer posttests (using the same IQ measure) to all children after the school year. We find that the enrichment training seems to have been associated with a gain in IQ for the initially low-scoring children; further, the high-scoring children appear to have lost IQ points; finally, the average children are unchanged with respect to IQ. The claim is often made that such findings are mere reflections of the regression effects described above, in which scores have a tendency to change on posttest in the direction of the mean.

The reader must evaluate this sort of claim, or any claim that real change is measured by pretests and posttests, in the light of the magnitude of change expected on the basis of regression. For example, if r_{XY} were .8, then we would expect regression effects on the order of magnitude of .2 of a standard deviation.[19] If the observed changes are very much larger than that in either direction, the possibility of real gain or loss should be entertained.

---

[19] Since here $z'_Y = .8 z'_X$ we would expect each predicted z'_Y to be $1 - .8$ or .2 less than its corresponding z_X. A coefficient of .8 represents a proportionate reduction of .2.

4

Analysis of Variance

Analysis of variance is a technique that allows us algebraically to divide the total variance, S^2, in a group of scores into portions. Under certain conditions (to be described) these portions of variance can serve as an unbiased estimate of the variation due to different identifiable sources, such as (*a*) different experimental treatments; (*b*) different identifiable subject characteristics; for example, sex, social class, age, IQ; (*c*) a combination of characteristics; or (*d*) unidentified subject characteristics. The arithmetic involved in analysis of variance can always be performed; but, the term "analysis of variance" should be reserved for the performance of one or more hypothesis tests using the portions of variance and a sampling distribution known as the F-distribution. In Section 4.1, let us proceed to the simplest situation in which analysis of variance is used. The more complex situations and their analyses will be built on this first case.

4.1 Rationale and Terms

In some formal experiments with one independent variable, more than just one experimental and one control situation may be needed. In Chapter 2, a study of the frustration-aggression hypothesis was illustrated (page 36). The experimental-group subjects in that study received some frustrating experience, whereas the control group subjects did not. Another investigator might feel that the independent variable, frustration, could better be characterized by more than two different situations. Perhaps he would retain the no-frustration control group and then develop three different experimental situations corresponding to low, moderate, and high amounts of frustration. Each of the four different experimental situations is called a *treatment*. Together, the four treatments—no, low, moderate, and high frustration—constitute an *operational definition* of frustration.

Since the experimental situations (that is, treatments) are created by the experimenter and are, as a result, literally controlled by him, the variable they define—frustration—is the *independent variable*. Aggression, which will be measured but not controlled by the experimenter, is the *dependent variable*. It might, for example, be represented by some score, Y, made by an observer of the subject during a post-treatment interview. In this example, the four particular frustration treatments are all that there are; that is, no sampling procedure is used to select them. Frustration, the independent variable, is therefore called a *fixed effect;* since there is only one independent variable, the analysis is called a *one-way* analysis of variance.

The one-way analysis of variance of this example will provide four treatment means. Six t-tests could be made comparing two of the means at a time. Analysis of variance is a technique for making these comparisons all at once with a single preset level of significance. The variance that is to be analyzed into parts is S_Y^2, the total variance for the dependent variable, aggression. Let us examine the expression for an unbiased estimate of S_Y^2 and the associated terminology.

$$S_Y^2 = \frac{\sum_{i=1}^{N} (Y_i - \bar{Y}_Y)^2}{N - 1} = \text{total sum of squares} = SS_{\text{total}} \qquad (4\text{-}1)$$

$$\phantom{S_Y^2 = \frac{\sum}{N-1}} = \text{total degrees of freedom} = \text{total } df$$

In Equation 4-1, $\bar{Y}$ indicates the mean of all the scores. The phrase "total df" refers to the number of scores independently free to vary. (See page 53.) The total sum of squares can be algebraically divided into two terms; one reflects variation *between* (that is, among) the four means; the other reflects the variation of individual scores *within* their own treatment groups. Here is what we have said:

Total sum of squares = *between* sum of squares
$$+ \ \textit{within} \ \text{sum of squares} \qquad (4\text{-}2)$$

or, in abbreviated form

$$SS_{total} = SS_{between} + SS_{within} \tag{4-3}$$

or, in algebraic formulae:

$$\sum_{j=1}^{4} \sum_{i=1}^{n_j} (Y_{ij} - \bar{Y})^2 = \sum_{j=1}^{4} n_j (\bar{Y}_j - \bar{Y})^2 \quad + \sum_{j=1}^{4} \sum_{i=1}^{n_j} (Y_{ij} - \bar{Y}_j)^2 \tag{4-4a}$$

Total sum of squares   Between sum of squares   Within sum of squares

Notice that a double summation sign ($\Sigma\Sigma$) is introduced here in the second term in Equation 4-4a. This is a notation that permits us to keep separate track of a score's identification with a particular subject, i, and a particular treatment group, j, of size n_j and with mean $\bar{Y}_j$. The double summation indicates that the algebraic expressions symbolized to its right are to be summed first within each treatment group and that the four sums so produced will then themselves be summed. This process is illustrated in Equation 4-4b, which is an extension of Equation 4-4a:

$$\sum_{j=1}^{4} \sum_{i=1}^{n_j} (Y_{ij} - \bar{Y})^2 =$$

$$\underbrace{n_1(\bar{Y}_1 - \bar{Y})^2 + n_2(\bar{Y}_2 - \bar{Y})^2 + n_3(\bar{Y}_3 - \bar{Y})^2 + n_4(\bar{Y}_4 - \bar{Y})^2}_{\text{Between sum of squares}} \tag{4-4b}$$

$$\underbrace{+ \sum_{i=1}^{n_1} (Y_{i1} - \bar{Y}_1)^2 + \sum_{i=1}^{n_2} (Y_{i2} - \bar{Y}_2)^2 + \sum_{i=1}^{n_3} (Y_{i3} - \bar{Y}_3)^2 + \sum_{i=1}^{n_4} (Y_{i4} - \bar{Y}_4)^2}_{\text{Within sum of squares}}$$

An estimate of variance could be made from the *within sum of squares* divided by its total *df*, $n_1 - 1 + n_2 - 1 + n_3 - 1 + n_4 - 1$ which totals to $N - 4$, or, in general, $N - k$. The estimate of variance so calculated is called the "mean square-within" and is found as

$$MS_W = \frac{SS_{within}}{N - k} \tag{4-5}$$

The estimate of variance based on the means is called the "mean square-between" and is found as

$$MS_{between} = \frac{SS_{between}}{k - 1} \tag{4-6}$$

The degrees of freedom for estimating the variation of k means is $k - 1$. It has been correctly implied that the two sums of squares add up to SS_{total} and that the two df's add up to $N - 1$, the total df. But the variance estimates do not add up to S_Y^2. The arithmetic in this section is summarized in Table 4.1.

4.1.1 The Hypothesis

What we really are after here is a method of testing the null hypothesis,

H_0: There is *no* relationship between frustration (independent variable) and aggression (dependent variable)

Here is the rationale. If H_0 is true, the population means, μ_1, μ_2, μ_3, μ_4, of the four treatment groups will be identical. In that case such slight variation among group (sample) means $\bar{Y}_1$, $\bar{Y}_2$, $\bar{Y}_3$, $\bar{Y}_4$ as does appear should be *no greater than* the variation among individual scores within the groups. In algebraic symbols H_0 states:

$$\mu_1 = \mu_2 = \mu_3 = \mu_4 = \mu, \text{ a constant} \tag{4-7}$$

and this has been shown to be equivalent to an assertion:

$$H_0: \qquad \sigma^2_{\text{between}} = \sigma^2_{\text{within}} \tag{4-8}$$

What are the alternatives to H_0? Any one of the following relations would contradict H_0:

$$\mu_1 < \mu_2 = \mu_3 = \mu_4$$

or

$$\mu_1 = \mu_2 < \mu_3 < \mu_4$$

or

$$\mu_1 = \mu_2 = \mu_3 < \mu_4$$

or

$$\mu_1 < \mu_2 < \mu_3 < \mu_4$$

Indeed, there are 31 logically possible order and/or equality relations among four means. Only one of these is the null hypothesis (Equation 4-7). Thus a rejection of H_0 implies only that *one* of the remaining thirty possibilities is true. It can be shown to follow, that if H_0 is false, then

$$\sigma^2_{\text{between}} > \sigma^2_{\text{within}} \tag{4-9}$$

which is a formulation of H_1, the alternative hypothesis. From Equa-

Table 4.1 *Analysis of variance calculations for a one-way, fixed-effects model. Test of the frustration-aggression hypothesis*

Frustration Treatment			
No	*Low*	*Moderate*	*High*
$Y_{11} = 0$	$Y_{12} = 1$	$Y_{13} = 3$	$Y_{14} = 3$
$Y_{21} = 1$	$Y_{22} = 1$	$Y_{23} = 2$	$Y_{24} = 4$
$Y_{31} = 2$	$Y_{32} = 2$	$Y_{33} = 3$	$Y_{34} = 3$
$Y_{41} = 1$	$Y_{42} = 2$	$Y_{43} = 2$	$Y_{44} = 2$
$Y_{51} = 1$	$Y_{52} = 0$	$Y_{53} = 1$	$Y_{54} = 3$

Sums

$$\sum_{i=1}^{5} Y_{i1} = 5 \qquad \sum_{i=1}^{5} Y_{i2} = 6 \qquad \sum_{i=1}^{5} Y_{i3} = 11 \qquad \sum_{i=1}^{5} Y_{i4} = 15$$

$$\sum_{i=1}^{5} \sum_{j=1}^{4} Y_{ij} = 37$$

Means

$$\bar{Y}_1 = 1.0 \qquad \bar{Y}_2 = 1.2 \qquad \bar{Y}_3 = 2.2 \qquad \bar{Y}_4 = 3.0$$

Notation key

Y_{ij} = Score on aggression of ith person in the jth treatment condition. Y is the dependent variable.

$\sum_{i=1}^{5} Y_{ij}$ = Sum of scores in jth treatment.

$\bar{Y}_j$ = Mean of the scores in jth treatment.
$\bar{Y}$ = Mean of all scores.
N = Total number of scores.
n_j = Number of scores in the jth treatment. In this case, all n_j are equal to 5.

Calculations

1. By the usual formula

$$SS_{\text{total}} = \sum_{i=1}^{n_j} \sum_{j=1}^{k} Y_{ij}^2 - \left(\sum_{i=1}^{n_j} \sum_{j=1}^{k} Y_{ij} \right)^2 / N$$

and

$$\begin{aligned}
\sum_{i=1}^{n_j} \sum_{j=1}^{k} Y_{ij}^2 &= 0^2 + 1^2 + 2^2 + 1^2 + 1^2 + 1^2 + 1^2 + 2^2 + 2^2 + 0^2 \\
&\quad 3^2 + 2^2 + 3^2 + 2^2 + 1^2 + 3^2 + 4^2 + 3^2 + 2^2 + 3^2 \\
&= 0 + 1 + 4 + 1 + 1 + 1 + 1 + 4 + 4 + 0 \\
&\quad 9 + 4 + 9 + 4 + 1 + 9 + 16 + 9 + 4 + 9 \\
&= 91
\end{aligned}$$

Table 4.1 (Continued) 93

and

$$\left(\sum_{i=1}^{n_j} \sum_{j=1}^{k} Y_{ij}\right)^2 = (37)^2 = 1369$$

Thus

$$SS_{\text{total}} = 91 - 1369/20 = 91 - 68.45$$
$$= 22.55$$

2. For illustrative purposes, SS_{within} is calculated from the definition Equation 4-4:

$$SS_{\text{within}} = \sum_{i=1}^{n_j} \sum_{i=1}^{k} (Y_{ij} - \bar{Y}_j)^2 = \sum_{j=1}^{k} SS_{\text{within group } j}$$

$$= SS_{\text{within 1}} + SS_{\text{within 2}} + SS_{\text{within 3}} + SS_{\text{within 4}}$$

and

$$\begin{aligned}
SS_{\text{within 1}} &= (0.0 - 1.0)^2 + (1.0 - 1.0)^2 + (2.0 - 1.0)^2 + (1.0 - 1.0)^2 + (1.0 - 1.0)^2 \\
&= (-1.00)^2 + (0.0)^2 + (1.0)^2 + (0.0)^2 + (0.0)^2 \\
&= 1.0 + 0.0 + 1.0 + 0.0 + 0.0 = \mathbf{2.00}
\end{aligned}$$

$$\begin{aligned}
SS_{\text{within 2}} &= (1.0 - 1.2)^2 + (1.0 - 1.2)^2 + (2.0 - 1.2)^2 + (2.0 - 1.2)^2 + (0.0 - 1.2)^2 \\
&= (-.2)^2 + (-.2)^2 + (.8)^2 + (.8)^2 + (-1.2)^2 \\
&= .04 + .04 + .64 + .64 + 1.44 = \mathbf{2.80}
\end{aligned}$$

$$\begin{aligned}
SS_{\text{within 3}} &= (3 - 2.2)^2 + (2 - 2.2)^2 + (3 - 2.2)^2 + (2 - 2.2)^2 + (1 - 2.2)^2 \\
&= (.8)^2 + (-.2)^2 + (.8)^2 + (-.2)^2 + (-1.2)^2 \\
&= .64 + .04 + .64 + .04 + 1.44 = \mathbf{2.80}
\end{aligned}$$

$$\begin{aligned}
SS_{\text{within 4}} &= (3.0 - 3.0)^2 + (4.0 - 3.0)^2 + (3.0 - 3.0)^2 + (2.0 - 3.0)^2 + (3.0 - 3.0)^2 \\
&= (0.0)^2 + (1.0)^2 + (0.0)^2 + (-1.0)^2 + (0.0)^2 \\
&= 0.0 + 1.0 + 0.0 + 1.0 + 0.0 = \mathbf{2.00}
\end{aligned}$$

Thus

$$SS_{\text{within}} = 2.00 + 2.80 + 2.80 + 2.00 = 9.60$$
$$MS_{\text{within}} = 9.60 \div 16 = .600$$

3. And from the definition Equation 4-3, SS_{between} is found:

$$SS_{\text{between}} = \sum_{j=1}^{k} n_j (\bar{Y}_j - \bar{Y})^2$$

$$\begin{aligned}
&= 5(1.00 - 1.85)^2 + 5(1.20 - 1.85)^2 + 5(2.20 - 1.85)^2 + (3.00 - 1.85)^2 \\
&= 5[(-.85)^2 + (-.65)^2 + (.35)^2 + (1.15)^2] \\
&= 5[.7225 + .4225 + .1225 + 1.3225] \\
&= 5 \times 2.5900 = 12.95
\end{aligned}$$

$SS_{\text{between}} = 12.95$, and from Equation 4-6, MS_{between} is found:

$$MS_{\text{between}} = 12.95 \div 3 = 4.319$$

4. Check: $SS_{\text{within}} + SS_{\text{between}} = 9.60 + 12.95 = 22.55 = SS_{\text{total}}$

tions 4-8 and 4-9 together, we can restate the hypotheses as

$$H_0: \quad \sigma^2_{\text{between}}/\sigma^2_{\text{within}} = 1 \qquad (4\text{-}8^*)$$

$$H_1: \quad \sigma^2_{\text{between}}/\sigma^2_{\text{within}} > 1 \qquad (4\text{-}9^*)$$

These expressions were obtained when both sides of the two Equations 4-8 and 4-9 were divided by σ^2_{within}. Notice that the ratio of the two variances cannot be less than 1.00 under either H_0 or H_1.

The null hypothesis, H_0, is tested by estimating the relevant ratio as

$$F = \widehat{S}^2_{\text{between}}/\widehat{S}^2_{\text{within}}$$

$$\text{(an estimate of } \sigma^2_{\text{between}}/\sigma^2_{\text{within}}) \qquad (4\text{-}10)$$

and referring F to its sampling distribution. The appropriate family of sampling distributions has been derived, and they have known mathematical equations from which the probability of an F *at least as large* as the one observed may be calculated. There is a distinct F-distribution for every distinctly different pair of numerator df and denominator df. The minimum values of F that are required for a preset level of significance and for various combinations of df's are available in table form. A part of such a table is reproduced as Table 4.2.

The null hypothesis for analysis of variance is thus tested according to the general logic outlined in Chapter 2. That is, several assump-

Table 4.2   *Minimum F-values* beyond $\alpha = .05$ for selected df*

Denominator df's	Numerator df's									
	1	*2*	*3*	*5*	*7*	*10*	*20*	*30*	*120*	*∞*
1	161.4	199.5	215.7	230.2	236.8	241.9	248.0	250.1	253.3	254.3
9	5.12	4.26	3.86	3.48	3.29	3.14	2.94	2.86	2.75	2.71
16	4.49	3.63	3.24	2.85	2.66	2.49	2.28	2.19	2.06	2.01
20	4.35	3.49	3.10	2.71	2.51	2.35	2.12	2.04	1.90	1.84
24	4.26	3.40	3.01	2.62	2.42	2.25	2.03	1.94	1.79	1.73
30	4.17	3.32	2.92	2.53	2.33	2.16	1.93	1.84	1.68	1.62
40	4.08	3.23	2.84	2.45	2.25	2.08	1.84	1.74	1.58	1.51
60	4.00	3.15	2.76	2.37	2.17	1.99	1.75	1.65	1.47	1.39
120	3.92	3.07	2.68	2.29	2.09	1.91	1.66	1.55	1.35	1.25
∞	3.84	3.00	2.60	2.21	2.01	1.83	1.57	1.46	1.22	1.00

*After Hays, W., *Statistics for Psychologists*. New York: Holt, Rinehart, and Winston, 1963, Table IV, p. 677.

tions are made, namely:

1. H_0: $F = 1$.
2. $F = \widehat{S}^2_{\text{between}}/\widehat{S}^2_{\text{within}}$ is a good estimate of $\sigma^2_{\text{between}}/\sigma^2_{\text{within}}$.

This assumption is strictly true only in case $\sigma^2_1 = \sigma^2_2 = \cdots \sigma^2_k = \sigma^2$, that is, the population values of the group variances are equal.

3. F follows the F-distribution for numerator df, $k - 1$, and denominator df, $N - k$.

This assumption is strictly true only when the scores within each group are normally distributed with identical mean values.

Next, these assumptions are used to find the probability that an F at least as large as that observed could occur. If this probability is arbitrarily low, say less than .05, H_0 (assumption 1) is rejected. Otherwise, H_0 is accepted. This hypothesis-testing procedure is known as "an F-test,"[1] and is a procedure of choice only when assumptions 2 and 3 are true.[2]

4.1.2 Directionality of the F Test

It should be observed that since F's can never be negative, the question of a one-direction versus a two-direction test does not arise. Note that, whenever there are three or more means being compared, the question of the direction of difference does not make sense. Either the means are all equal or they are not. Any prediction that refers to *only two means* could be tested by a normal curve or t-test procedure.

---

[1]The family of F-distributions are the sampling distributions of ratios of two variance estimates. More generally, any variable will have an F-distribution which is formed as

$$X = \frac{A/k}{B/l}$$

where $A = a^2_1 + a^2_2 + \cdots + a^2_k$
and $B = b^2_1 + b^2_2 + b^2_3 + \cdots + b^2_l$
and $a_1, a_2, \ldots a_k$ and $b_1, b_2 \ldots b_l$
are independent, standardized, and normally distributed variables.

[2]Values of F are not seriously distorted when the scores have nonnormal distributions if the n_s in the groups are equal or proportionate. Nor are they seriously distorted if the variance in groups are not equal if the n_s are equal. These states of affairs are summarized by the sentence, "If the groups have equal n_s, the F-test is 'robust against' violations of the assumptions of normality and equal variance."

4.1.3 Errors

As always, the probability of Type I error for the F-test is the preset level of significance, α. This is the probability that the investigator will decide that at least one of the several differences between means is not zero when in fact all of them are zero. The F-test has a lower Type I error probability for the entire analysis than would obtain if one ignored the F-test and tested the same hypotheses using as many t-tests as were required to compare each mean with each other mean. In the F-test, α is approximately equal in value to the probability of at least one erroneous judgment. For n t-tests this probability is considerably larger. For example, consider a study of three means with the α-level set at .05. This is the probability of at least one error using the F procedure. There would be three t-tests $(\bar{Y}_1 - \bar{Y}_2, \bar{Y}_1 - \bar{Y}_3, \bar{Y}_2 - \bar{Y}_3)$, and they would not be independent. Thus, if each had an .05 level of significance, they would introduce a probability of at least one error exceeding .05. This discrepency in error rates is the principal reason why an F-test is to be preferred to multiple t-tests.

The Type II error probability $(1 - \beta)$ is a complex result of the size and number of real differences; the denominator degrees of freedom; and α, the probability of Type I error. The value of $1 - \beta$ can be calculated for any fixed alternative, H_1, to H_0. Since there are so very many of these, varying in both number and size of real differences, this is rarely done in practice. Mathematical statisticians have shown, however, that *when the assumptions* (2 and 3, page 95) *are met*, the F-test is *uniformly most powerful*. This means that the power β is higher and the probability $(1 - \beta)$ of Type II error smaller than for any other hypothesis test procedure.

4.1.4 Interpreting the Presentation of Results

The reader of journal articles will frequently encounter an analysis-of-variance summary such as Table 4.3.

Table 4.3 *Summary of the Analysis of variance of Table 4.1*

Source	df	SS	MS	f
Between groups (frustration)	$K - 1 = 3$	12.95	4.319	7.198*
Within groups (error)	$nK - K = 16$	9.60	.600	
TOTAL	$nK - 1 = 19$	22.55		

*Exceeds the tabled (Table 4.2) value for $\alpha = .05$ for df's of 3 and 16.

There are several terms in the table that require comment. The term "source" is meant to indicate that the factors in the source column are a list of mutually exclusive and exhaustive sources of variation in the dependent variable (aggression). One of these sources is labeled "error." This is the within-groups term. The error term in any analysis of variance refers to the variance "between" subjects remaining after all influence of experimental treatments has been removed. This remaining within-groups variance ("between" S's who are within treatments) represents individual differences and is an estimate of how much variation we can expect from *any* source even if that source is not a real effect. The label "error" then indicates that the term so labeled estimates the magnitude the other terms might reach merely by chance. The ratio, F, of any other source's variance to error variance would be 1.00 if only chance is operating. Thus the error (denominator) term for any F-ratio test is an estimate of the size of the numerator term in the ratio *if H_0 is true*.

The column next to Source in Table 4.3 is labeled "*df*" for degrees of freedom; the number of scores free to vary in calculating each variance term are shown in this column. Since S_Y^2 has $df = N - 1$, the *df* column must total $N - 1$ since the list of sources must be mutually exclusive and exhaustive; that is, the list must include all possible nonoverlapping sources, and they must add to the total. The next column, labeled "*SS*," is for the sums of squares

$$\sum_{i=1}^{k} (\bar{Y}_i - \bar{Y})^2 \quad \text{(between)}$$

and

$$\sum_{i=1}^{k} \sum_{j=1}^{n_i} (Y_{ij} - \bar{Y}_i)^2 \quad \text{(within)}$$

$$\sum_{i=1}^{k} \sum_{j=1}^{n_i} (Y_{ij} - \bar{Y})^2 \quad \text{(total)}$$

The next column in Table 4.3 is labeled "*MS*" for mean squares. These are the unbiased estimates of variance due to each source.

$$MS_{\text{between}} = \hat{S}_{\text{between}}^2 = \frac{SS_{\text{between}}}{k - 1}$$

$$MS_{\text{within}} = \hat{S}_{\text{within}}^2 = \frac{SS_{\text{within}}}{Nk - k}$$

Table 4.4   *Means and standard deviations* for the analysis of Table 4.1*

	Frustration treatment		
No	*Low*	*Moderate*	*High*
$\bar{Y}_1 = 1.00$	$\bar{Y}_2 = 1.20$	$\bar{Y}_3 = 2.20$	$\bar{Y}_4 = 3.00$
$S = .63$	$S = .75$	$S = .74$	$S = .63$

* In tables like this, it is customary to use S instead of $\hat{S}$ since the tables are for descriptive summary. The analyses leading to the F-tests use what amount to $\hat{S}$ estimates from the various sources.

$$MS_{\text{total}} = \hat{S}^2_{\text{total}} = \frac{SS_{\text{total}}}{N - 1}$$

The term "mean square" comes from the fact that these terms may indeed be thought of as means of squared errors, where an error is a deviation of a score or group mean from its mean. The final column contains F's; in the example analysis, there is only one F. This F is 7.198, indicating a significance of $\alpha = .05$.

Actually, all of the information you can justifiably use is contained in a table of means and standard deviations, such as Table 4.4. By inspection you can note that the two highest experimental means appear to differ from the lower experimental and control means but not from each other. You would need to check for an indication that there is at least one significant difference among four means. Such information may often be found indicated in the text by a parenthetical F-value with its associated p as $(F = 7.198, p < .05)$ or it may be indicated by a note next to the F in a table such as Table 4.3. The verification of the difference as being where it appears to be (between the two lowest and two highest groups) must be done by a test of multiple comparisons following a significant F. These tests are discussed in the following section.

4.2 Multiple Comparisons

When an F-test results in the rejection of H_0, the investigator may follow the test with one for simultaneous or multiple comparisons of the several means. These tests are designed to identify which of the several possible differences between means or combinations of means is significant.

The most commonly cited multiple-comparisons procedures are

those of Duncan, Tukey, Scheffe, and Neuman-Keuls. The Duncan multiple-range test has greater power (β) to reject H_0 than some of the other procedures. As a result, the Duncan test also has a significance level (α) higher than that set for the corresponding analysis of variance.

The Scheffe procedure is appropriate when (1) comparisons among means are to be made and (2) the investigator wishes to have the probability of *at least one* Type I error in all possible comparisons held to $\alpha = .05$ (or some other preset value). The test is essentially the multiplication of the minimum F-value for the α-level used (symbolized F_α) by its numerator degrees of freedom (which may be symbolized $df_{\text{bet.}}$). Before significance is claimed, the absolute value of any difference between means divided by the error term of the F is required to exceed the value of $\sqrt{df_{\text{bet.}}F_\alpha}$. In the example of Table 4.1, this critical value is found as $V = \sqrt{3 \times 3.24} = \sqrt{9.72} = 3.12$. Thus only the differences between the means of the high- and no-frustration groups differ significantly. All other pairwise comparisons are nonsignificant, although some complex comparison involving three or more of the means might be significant.

The Neuman-Keuls procedure takes into account the order of the means and is appropriate when that order has been predicted or has importance for interpretation. The number of ordered steps that separate one means from another taken into account in making the test: an ordered step being a difference in ordinal position. For instance, the first-ranked and third-ranked means are two ordered steps apart. A statistic called the studentized range is used to evaluate the magnitude of a difference between means that are a stated number of steps apart. The same size difference may be significant at one step but not at two steps. This procedure should only be used if the concept of ordered steps is meaningful for the study being done—a question the reader must address.

The Tukey procedure amounts to a series of t-tests with these changes: (1) The size of t required for significance is adjusted so that the probability of at least one Type I error is equal to the original α. (2) Also,

$$\sqrt{\frac{2\,MS_{\text{error}}}{N}}$$

is used as

$$\widehat{S}_{\bar{Y}_1 - \bar{Y}_2}$$

for all comparisons, in contrast to an estimate based on only two groups, which would be used with a conventional t-test. If the equal-variance assumption is correct, this estimate is better than one based on only two groups because more relevant data are used.

The reader will want to interpret any discussion by an author of specific difference between two (of several) means in light of the multiple-comparisons procedure used. If multiple *t*-tests are employed instead of a multiple-comparisons test, Type I error probability will exceed α. Also, if the Duncan test is used inappropriately, the Type I error probability will be increased. The Neuman-Kuels procedure may actually reduce the probability of Type I error if it is used with an ordering that is really not meaningful; in this sense it is the most conservative test.

4.3 Two-Way Analyses of Variance

4.3.1 Factorial Design

When two or more independent variables are simultaneously manipulated, there are several combinations of treatments possible. For example, if the first independent variable has four treatments and the second has three treatments, there are a total of four times three, or 12 treatment combinations; when each of these combinations is administered to several subjects, we have what is known as a factorial experiment.[3] In descriptions of statistical analyses, the independent variables often are referred to as "factors," and the treatments within a factor are called "levels." To illustrate, let us reconsider our experiment with frustration (first factor) and aggression (dependent variable). We will add a second independent variable: the subject's tendency to verbal aggression as scored by his acquaintances prior to the start of the frustration experiment. Let's suppose we select equal numbers of subjects who are high, average, and low on verbal aggression according to peer ratings. Our final experimental design might be the one shown, along with hypothetical data, in Table 4.5. An author might refer to this table as a "three-by-four factorial design with five subjects per cell": a cell here being a combination of a single level on one factor with a single level on another factor. (For example, one cell is defined by the combination of No Frustration and High Verbal Aggression.)

4.3.1.1 Comparison to a one-way analysis. Notice that Table 4.5 could be broken down into three rows of four cells each. Indeed, we

---

[3]The term factorial experiment is extended by some authors to include designs in which the treatments on one of the independent variables are all administered to each subject and those on the other to different subjects.

could depict the design as a 12-group *one*-way design in which our 12 groups are defined in terms of combinations of levels on our two factors; this conceptualization is shown in Table 4.6. The total variance would be identical in Tables 4.5 and 4.6 and *so would the variances for between cells (groups) and within cells*. An *F*-test of the cells' effect would produce the same *F* in both cases; indeed, the arithmetical procedures would be identical.

The difference is this: in the three-by-four factorial design of Table 4.5, if we find a significant *F* for "between cells," we will still be uncertain about whether it arose mainly from differences between cells from different rows or between cells from different columns, or from some complex pattern of row-and-column combinations. Complex multiple-comparisons procedures could be devised to contrast the different rows, the different columns, and the cells that differ both in row and column.

Fortunately, there is an equivalent but easier way. The total sum of squares for Table 4.5 can be partitioned into parts to show row–column interaction (to be discussed) and "within cells" (error). The row, column, and interaction sums of squares from Table 4.5 total to the "between-groups" sum of squares in a one-way design such as that of Table 4.6. The "within-cells" sum of squares is the same in both tables. This can be seen by comparing the algebraic partitioning of SS_{total} in Tables 4.7a and 4.7b.

Tables 4.7a and 4.7b present sample arithmetic calculations for the hypothetical expressions of Tables 4.5 and 4.6. For reference, Table 4.8 defines all notations that are used in algebraic expressions in two-way analyses of variance. Table 4.9a presents the summary of analysis of variance for Table 4.5, and Table 4.9c presents the means and standard deviations for the same analysis. Table 4.9b presents the summary of the analysis for Table 4.6.

The complicated algebra and arithemetic needed for the two-way design can be conceptualized as a performance in several steps of the simplest division involved in the one-way design. Every two-way analysis has a smallest unit, a cell, which includes the smallest group of scores derived under identical treatment conditions. If the number of scores in a cell exceeds one, then a mean square based the variation among scores within cells is the error term. Actually, it is an estimate of variance due to factors other than those represented in the experiment. If the number in a cell is one, then there can be no "within-cell" variation; variation in some other, larger, subsets of scores will be used as the error term. The choices of an error term rests on assumptions that are difficult for a reader to evaluate without access to the actual data and without considerable expertese. A reader can,

Table 4.5 *Three-by-four factorial design for the study of the frustration-aggression hypothesis*

Aggressiveness level	Frustration treatment				Row sums of Y and row sums of Y^2
	No	Low	Moderate	High	
HIGH	$Y_{111} = 2$	$Y_{112} = 4$	$Y_{113} = 4$	$Y_{114} = 4$	
	$Y_{211} = 3$	$Y_{212} = 3$	$Y_{213} = 3$	$Y_{214} = 4$	
	$Y_{311} = 3$	$Y_{312} = 4$	$Y_{313} = 2$	$Y_{314} = 3$	
	$Y_{411} = 2$	$Y_{412} = 2$	$Y_{413} = 4$	$Y_{414} = 3$	
	$Y_{511} = 4$	$Y_{512} = 2$	$Y_{513} = 3$	$Y_{514} = 2$	
Column sums of Y within row 1	$\sum\limits_{i=1}^{5} Y_{i11} = 14$	$\sum\limits_{i=1}^{5} Y_{i12} = 15$	$\sum\limits_{i=1}^{5} Y_{i13} = 16$	$\sum\limits_{i=1}^{5} Y_{i14} = 16$	$\sum\limits_{i=1}^{4}\sum\limits_{k=1}^{} Y_{i1k} = 61$
Column sums of Y^2 within row 1	$\sum\limits_{i=1}^{5} Y_{i11}^2 = 42$	$\sum\limits_{i=1}^{5} Y_{i12}^2 = 49$	$\sum\limits_{i=1}^{5} Y_{i13}^2 = 54$	$\sum\limits_{i=1}^{5} Y_{i14}^2 = 54$	$\sum\limits_{i=1}^{4}\sum\limits_{k=1}^{} Y_{i1k}^2 = 199$
MODERATE	$Y_{121} = 0$	$Y_{122} = 1$	$Y_{123} = 1$	$Y_{124} = 4$	
	$Y_{221} = 0$	$Y_{222} = 2$	$Y_{223} = 2$	$Y_{224} = 4$	
	$Y_{321} = 2$	$Y_{322} = 2$	$Y_{323} = 3$	$Y_{324} = 4$	
	$Y_{421} = 1$	$Y_{422} = 1$	$Y_{423} = 3$	$Y_{424} = 1$	
	$Y_{521} = 1$	$Y_{522} = 1$	$Y_{523} = 4$	$Y_{524} = 4$	
Column sums of Y within row 2	$\sum\limits_{i=1}^{5} Y_{i21} = 4$	$\sum\limits_{i=1}^{5} Y_{i22} = 7$	$\sum\limits_{i=1}^{5} Y_{i23} = 13$	$\sum\limits_{i=1}^{5} Y_{i24} = 17$	$\sum\limits_{i=1}^{5}\sum\limits_{k=1}^{4} Y_{i2k} = 41$
Column sums of Y^2 within row 2	$\sum\limits_{i=1}^{5} Y_{i21}^2 = 6$	$\sum\limits_{i=1}^{5} Y_{i22}^2 = 11$	$\sum\limits_{i=1}^{5} Y_{i23}^2 = 39$	$\sum\limits_{i=1}^{5} Y_{i24}^2 = 65$	$\sum\limits_{i=1}^{5}\sum\limits_{k=1}^{4} Y_{i2k}^2 = 121$

(Continued)

Table 4.5 (Continued)

LOW			
$Y_{131} = 1$	$Y_{132} = 2$	$Y_{133} = 0$	$Y_{134} = 0$
$Y_{231} = 2$	$Y_{232} = 2$	$Y_{233} = 0$	$Y_{234} = 1$
$Y_{331} = 1$	$Y_{332} = 1$	$Y_{333} = 1$	$Y_{334} = 0$
$Y_{431} = 2$	$Y_{432} = 1$	$Y_{433} = 2$	$Y_{434} = 1$
$Y_{531} = 2$	$Y_{532} = 1$	$Y_{533} = 1$	$Y_{534} = 0$

Column sums of Y within row 3

$\sum_{i=1}^{5} Y_{i31} = 8$	$\sum_{i=1}^{5} Y_{i32} = 7$	$\sum_{i=1}^{5} Y_{i33} = 4$	$\sum_{i=1}^{5} Y_{i34} = 2$	$\sum_{i=1}^{5}\sum_{k=1}^{4} Y_{i3k} = 21$

Column sums of Y^2 within row 3

$\sum_{i=1}^{5} Y_{i31}^2 = 14$	$\sum_{i=1}^{5} Y_{i32}^2 = 11$	$\sum_{i=1}^{5} Y_{i33}^2 = 6$	$\sum_{i=1}^{5} Y_{i34}^2 = 2$	$\sum_{i=1}^{5}\sum_{k=1}^{4} Y_{i3k}^2 = 33$

TOTAL

Column sums of Y

$\sum_{i=1}^{5}\sum_{j=1}^{3} Y_{ijk} = 26$	$\sum_{i=1}^{5}\sum_{j=1}^{3} Y_{ijk} = 29$	$\sum_{i=1}^{5}\sum_{j=1}^{3} Y_{ijk} = 33$	$\sum_{i=1}^{5}\sum_{j=1}^{3} Y_{ijk} = 35$

Column sums of Y^2

$\sum_{i=1}^{5}\sum_{j=1}^{3} Y_{ijk}^2 = 62$	$\sum_{i=1}^{5}\sum_{j=1}^{3} Y_{ijk}^2 = 71$	$\sum_{i=1}^{5}\sum_{j=1}^{3} Y_{ijk}^2 = 99$	$\sum_{i=1}^{5}\sum_{j=1}^{3} Y_{ijk}^2 = 121$

GRAND SUMS
$$\sum_{k=1}^{4}\sum_{j=1}^{3}\sum_{i=1}^{5} Y_{ijk} = 123$$
$$\sum_{k=1}^{4}\sum_{j=1}^{3}\sum_{i=1}^{5} Y_{ijk}^2 = 353$$

$$N = 60$$

Table 4.6 *One-way analysis of variance design corresponding to the factorial design of Table 4.5*

	Row 1 of Table 4.5				Row 2 of Table 4.5				Row 3 of Table 4.5				Total
	No Frust.; High Agg.*	Low Frust.; High Agg.	Mod. Frust.; High Agg.	High Frust.; High Agg.	No Frust.; Mod. Agg.	Low Frust.; Mod. Agg.	Mod. Frust.; Mod. Agg.	High Frust.; Mod. Agg.	No Frust.; Low Agg.	Low Frust.; Low Agg.	Mod. Frust.; Low Agg.	High Frust.; Low Agg.	
	$Y_{11}=2$	$Y_{12}=4$	$Y_{13}=4$	$Y_{14}=4$	$Y_{15}=0$	$Y_{16}=1$	$Y_{17}=1$	$Y_{18}=4$	$Y_{19}=1$	$Y_{1,10}=2$	$Y_{1,11}=0$	$Y_{1,12}=0$	$\sum_{i=1}^{5}\sum_{j=1}^{12} Y_{ij}=123$
	$Y_{21}=3$	$Y_{22}=3$	$Y_{23}=3$	$Y_{24}=0$	$Y_{25}=0$	$Y_{26}=2$	$Y_{27}=2$	$Y_{28}=4$	$Y_{29}=2$	$Y_{2,10}=2$	$Y_{2,11}=0$	$Y_{2,12}=1$	
	$Y_{31}=3$	$Y_{32}=4$	$Y_{33}=2$	$Y_{34}=3$	$Y_{35}=2$	$Y_{36}=2$	$Y_{37}=3$	$Y_{38}=4$	$Y_{39}=2$	$Y_{3,10}=1$	$Y_{3,11}=1$	$Y_{3,12}=0$	$\sum_{i=1}^{5}\sum_{j=1}^{12} Y_{ij}^2=353$
	$Y_{41}=2$	$Y_{42}=2$	$Y_{43}=4$	$Y_{44}=3$	$Y_{45}=1$	$Y_{46}=1$	$Y_{47}=3$	$Y_{48}=4$	$Y_{49}=2$	$Y_{4,10}=1$	$Y_{4,11}=1$	$Y_{4,12}=1$	$\bar{Y}=2.05$
	$Y_{51}=4$	$Y_{52}=2$	$Y_{53}=3$	$Y_{54}=2$	$Y_{55}=1$	$Y_{56}=1$	$Y_{57}=4$	$Y_{58}=4$	$Y_{59}=2$	$Y_{5,10}=1$	$Y_{5,11}=1$	$Y_{5,12}=0$	
Sums of Y	$\sum_{i=1}^{5} Y_{i1}=14$	$\sum_{i=1}^{5} Y_{i2}=15$	$\sum_{i=1}^{5} Y_{i3}=16$	$\sum_{i=1}^{5} Y_{i4}=16$	$\sum_{i=1}^{5} Y_{i5}=4$	$\sum_{i=1}^{5} Y_{i6}=7$	$\sum_{i=1}^{5} Y_{i7}=13$	$\sum_{i=1}^{5} Y_{i8}=17$	$\sum_{i=1}^{5} Y_{i9}=8$	$\sum_{i=1}^{5} Y_{i10}=7$	$\sum_{i=1}^{5} Y_{i11}=4$	$\sum_{i=1}^{5} Y_{i12}=2$	
Sums of Y^2	$\sum_{i=1}^{5} Y_{i1}^2=42$	$\sum_{i=1}^{5} Y_{i2}^2=49$	$\sum_{i=1}^{5} Y_{i3}^2=54$	$\sum_{i=1}^{5} Y_{i4}^2=54$	$\sum_{i=1}^{5} Y_{i5}^2=6$	$\sum_{i=1}^{5} Y_{i6}^2=11$	$\sum_{i=1}^{5} Y_{i7}^2=39$	$\sum_{i=1}^{5} Y_{i8}^2=65$	$\sum_{i=1}^{5} Y_{i9}^2=14$	$\sum_{i=1}^{5} Y_{i10}^2=11$	$\sum_{i=1}^{5} Y_{i11}^2=6$	$\sum_{i=1}^{5} Y_{i12}^2=2$	
Means ($\bar{Y}_j$)	2.8	3.0	3.2	3.2	0.8	1.4	2.6	3.4	1.6	1.4	0.8	0.4	

*This group corresponds to a cell in Table 4.5. Similarly, so do other groups.

Table 4.7a *Formulas and arithmetic for analysis of Table 4.5*

$$SS_{total} = \sum_{i=1}^{5}\sum_{j=1}^{3}\sum_{k=1}^{4} Y_{ijk}^2 - \frac{1}{N}\left[\sum_{i=1}^{5}\sum_{j=1}^{3}\sum_{k=1}^{4} Y_{ijk}\right]^2 = 353 - 252.15 = \mathbf{100.85}$$

$$SS_{bet.\,cells} = \frac{1}{5}\sum_{j=1}^{3}\sum_{k=1}^{4}\left(\sum_{i=1}^{5} Y_{ijk}\right)^2 - \underset{G}{\underbrace{}} = \frac{1}{5}[(14)^2 + (15)^2 + (16)^2 + (16)^2$$
$$+ (4)^2 + (7)^2 + (13)^2 + (17)^2 + (8)^2 + (7)^2 + (4)^2 + (2)^2] - G$$
$$= \frac{1}{5}(1589) - 252.15 = \mathbf{65.65}$$

$$SS_{within\,cells} = \sum_{j=1}^{3}\sum_{k=1}^{4}\left[\sum_{i=1}^{5} Y_{ijk}^2 - \frac{1}{5}\left(\sum_{i=1}^{5} Y_{ijk}\right)^2\right] = \left[42 - \frac{1}{5}(14)^2\right]$$
$$+ \left[49 - \frac{1}{5}(15)^2\right] + \left[54 - \frac{1}{5}(16)^2\right] + \left[6 - \frac{1}{5}(4)^2\right]$$
$$+ \left[11 - \frac{1}{5}(7)^2\right] + \left[39 - \frac{1}{5}(13)^2\right] + \left[65 - \frac{1}{5}(17)^2\right] + \left[14 - \frac{1}{5}(8)^2\right]$$
$$+ \left[11 - \frac{1}{5}(7)^2\right] + \left[6 - \frac{1}{5}(4)^2\right] + \left[2 - \frac{1}{5}(2)^2\right]$$
$$= 2.8 + 4.0 + 2.8 + 2.8 + 1.2 + 5.2 + 7.2 + 1.2$$
$$+ 2.8 + 1.2 = \mathbf{35.20}$$

$$SS_{rows} = \frac{1}{20}\sum_{k=1}^{3}\left[\sum_{j=1}^{4}\sum_{i=1}^{5} Y_{ijk}\right]^2 - G = \frac{1}{20}[(61)^2 + (41)^2 + (21)^2] - 252.15 = \mathbf{40.00}$$

$$SS_{columns} = \frac{1}{15}\sum_{j=1}^{4}\left[\sum_{k=1}^{3}\sum_{i=1}^{5} Y_{ijk}\right]^2 - G = \frac{1}{15}[(26)^2 + (29)^2 + (33)^2 + (35)^2]$$
$$- 252.15 = \mathbf{3.25}$$

$$SS_{interaction} = SS_{bet.\,cells} - SS_{rows} - SS_{cols.} = 65.65 - 40.00 - 3.25 = \mathbf{22.40}$$

Table 4.7b *Formulas and arithmetic for analysis of Table 4.6*

$$SS_{bet.\,groups} = \frac{1}{5}\sum_{j=1}^{12}\left[\sum_{i=1}^{5} Y_{ij}\right]^2 - \underset{G}{\underbrace{\frac{1}{N}\left[\sum_{i=1}^{5}\sum_{i=1}^{5} Y_{ij}\right]^2}}$$

$$= \frac{1}{5}[(14)^2 + (15)^2 + (16)^2 + (16)^2 + (4)^2 + (7)^2 + (13)^2 + (17)^2 + (8)^2$$
$$+ (7)^2 + (4)^2 + (2)^2] - \frac{1}{60}(123) = \frac{1}{5}(1589) - 252.15 = \mathbf{65.65},\ which\ is$$
$$SS_{bet.\,cells}\ in\ Table\ 4.7a.$$

$$SS_{within\,groups} = \sum_{j=1}^{12}\left[\sum_{i=1}^{5} Y_{ij}^2 - \frac{1}{5}\left(\sum_{i=1}^{5} Y_{ij}\right)^2\right]$$

$$= \left[42 - \frac{1}{5}(14)^2\right] + \left[49 - \frac{1}{5}(15)^2\right] + \left[54 - \frac{1}{5}(16)^2\right] + \left[54 - \frac{1}{5}(16)^2\right]$$
$$+ \left[6 - \frac{1}{5}(4)^2\right] + \left[11 - \frac{1}{5}(7)^2\right] + \left[39 - \frac{1}{5}(13)^2\right] + \left[65 - \frac{1}{5}(17)^2\right]$$
$$+ \left[14 - \frac{1}{5}(8)^2\right] + \left[11 - \frac{1}{5}(7)^2\right] + \left[6 - \frac{1}{5}(4)^2\right] + \left[2 - \frac{1}{5}(2)^2\right]$$
$$= 2.8 + 4.0 + 2.8 + 2.8 + 1.2 + 5.2 + 7.2 + 1.2 + 1.2 + 2.8 + 1.2$$
$$= \mathbf{35.20},\ which\ is\ SS_{within\,cells}\ in\ Table\ 4.7a.$$

$$SS_{total} = \sum_{j=1}^{12}\sum_{i=1}^{5} Y_{ij}^2 - G = 353 - 252.15 = \mathbf{100.85},\ which\ is\ also\ SS_{total}$$
$$in\ Table\ 4.7a.\ The\ total\ variance\ is\ identical\ for\ the\ two\ analyses.$$

Table 4.8 *Notation for two-way analysis of variance in Table 4.5*

Y_{ijk} = The ith score in the jth row and the kth column. Thus Y_{234} is the second score in the cell in row 3, column 4. $Y_{234} = 1$.

$$\sum_{i=1}^{5} \sum_{j=1}^{3} \sum_{k=1}^{4} Y_{ijk} = \text{The sum of all the scores.}$$

$$\sum_{i=1}^{5} \sum_{j=1}^{3} Y_{ijk} = \text{The sum of all the scores in the } k\text{th column. Thus}$$

$$\sum_{i=1}^{5} \sum_{i=1}^{3} Y_{ij2} = \text{The sum of the scores in column 2.}$$

$$\sum_{i=1}^{5} \sum_{i=1}^{3} Y_{ij2} = 7$$

$$\sum_{i=1}^{5} \sum_{k=1}^{4} Y_{ijk} = \text{The sum of all the scores in the } j\text{th row. Thus}$$

$$\sum_{i=1}^{5} \sum_{k=1}^{4} Y_{i1k} = \text{The sum of the scores in row 1.}$$

$$\sum_{i=1}^{5} \sum_{k=1}^{4} Y_{i1k} = 61$$

$$\sum_{i=1}^{5} Y_{ijk} = \text{The sum of all the scores in cell } jk. \text{ Thus}$$

$$\sum_{i=1}^{5} Y_{i23} = \text{The sum of the scores in cell 23.}$$

$$\sum_{i=1}^{5} Y_{i23} = 13$$

$$\left[\sum_{i=1}^{5} Y_{ijk} \right]^2 = \text{The squared sum of all the scores in cell } jk. \text{ Thus}$$

$$\left[\sum_{i=1}^{5} Y_{i23} \right]^2 = 13^2 = 169$$

$$\sum_{i=1}^{5} Y_{ijk}^2 = \text{The sum of the squared scores in cell } jk. \text{ Thus}$$

$$\sum_{i=1}^{5} Y_{i23}^2 = 39$$

N = Number of scores in the table.
n_{jk} = Number of scores in a cell.
n_j = Number of scores in a row.
n_k = Number of scores in a column.
$\bar{Y}_{jk}, \bar{Y}_j,$ and $\bar{Y}_k$ = the means of cells, rows, and columns, respectively.
$\bar{\bar{Y}}$ = The mean of all N scores.

Table 4.9a *Summary of the analysis of variance of Table 4.5*

Source	df	SS	MS	f
Aggressiveness (rows)	2	40.00	20.00	27.285*
Frustration (cols.)	3	3.25	1.083	1.478†
Interaction	6	22.40	3.733	5.093*
Within cells (error)	48	35.20	0.733	
TOTAL	59	100.85		

*Significant at $\alpha < .01$.   †Not significant.

Table 4.9b *Summary of the analysis of variance of Table 4.6*

Source	df	SS	MS	f
Treatments (between)	11	65.650	5.968	8.142*
Within groups (error)	48	35.200	0.733	

*Significant at $\alpha < .01$.

Table 4.9c *Means and standard deviations for Table 4.5*

Aggressiveness		Frustration treatments				Means of Rows
		No	Low	Mod.	High	
High	Mean	2.80	3.00	3.20	3.20	
	SD	.746	.894	.746	.746	3.05
Moderate	Mean	0.80	1.40	2.60	3.40	
	SD	.746	.490	1.020	1.204	2.05
Low	Mean	1.60	1.40	0.80	0.40	
	SD	.490	.490	.746	.490	1.05
Means of cols.		1.733	1.933	2.200	2.333	$2.05 = \bar{Y}$

however, use this guideline: the assumptions of equal variance and normal distribution for scores apply to the populations that provided scores *within each of these smallest units,* be they groups (one-way, fixed effects), cells (two-way, factorial), or some other unit.

A table of standard deviations, such as Table 4.9c, can be used to spot gross deviations from the equal variance assumption;[4] non-normality is not quite so easily detected in summary statistics.

[4]Of course, a statistical test would have to be performed to warrant any claim that two or more variances really differ. There are procedures available for this purpose and the reader may encounter a legitimate claim that an investigator tested his data for "homogeneity of variance" as the null hypothesis of no differences is called.

Remember that all judgments about the distribution of scores must be based on statements about them made in the article or book, or else on logical considerations based on the definition of the score. It would be well for the reader to remember that analysis of variance is robust against—that is, still valid in spite of—moderate departures from the assumptions when the n's in these smallest units are equal (for example, when the n_{ij}'s are equal in a two-way factorial design).

If there is a marked deviation, as, for instance, when one cell in the factorial design has a much larger standard deviation than any of the others, you should be conservative in your interpretation of the reported results. You may also wish to scrutinize such a standard deviation, for it may in itself be an interesting fact. In our examples, moderately aggressive and moderate- and high-frustration subjects vary among themselves more than any other groups of subjects do. It would appear that a subject's reaction to this particular set of circumstances (treatment combination) is more idiosyncratic than that to any other. Previous knowledge about reactions to frustration could be used to suggest an explanation and new research conducted to test it.

4.3.1.2 Interpretation of main effects. An "effect" refers to differences between units of analysis. Thus, in discussing the two-way factorial experiment, we can speak of a row effect, a column effect, a "between-cell" effect, and a "within-cell" effect. The row and column effects are called "main effects" (and are part of the "between-cell" effect). "An effect" (singular), refers to the deviation of one statistic from the mean of all like statistics. For example, in Table 4.9c, the effect of row 1 is $(\bar{Y}_1 - \bar{Y})$. The reader is more likely to encounter the term "an effect" in the context of a discussion of all the effects from one source. Thus the investigator will report his results with such comments as, "There was no frustration effect, but the aggressiveness effect and the aggression-by-frustration treatments interaction (discussed later) were both significant ($F = 5.093$, $p < .01$; $F = 27.285$, $p < .01$, respectively)."

The last statement above is a report of results from the analysis of variance of Table 4.7a, which is summarized in Tables 4.9a and 4.9c. Note that in Tables 4.5, 4.7a, 4.7b, and 4.9c, the aggression variable is defined by rows and the frustration treatments are shown in columns. Thus "a significant aggression effect" implies that the population means corresponding to the three rows μ_1, μ_2, and μ_3, and thus the effects $(\mu_1 - \mu)$, $(\mu_2 - \mu)$, and $(\mu_3 - \mu)$ differ from each other. This difference is reflected in the fact that an estimate of the mean of the squares—made using the appropriate *df*, 2—is large. That is, the mean square of these terms—these effects—is large. It is large enough that its ratio to mean square error produces an F large enough to cause the null hypothesis of no row differences to be rejected. The phrase "no column (frustration treatment) effect" can similarly be interpreted by noting that the four means, $\bar{Y}_1$, $\bar{Y}_2$, $\bar{Y}_3$, and $\bar{Y}_4$, do not differ enough to produce a large mean square. Hence F is near enough to its expected value that the null hyphothesis of no frustration effect may logically be accepted.

4.3.1.3 The concept of interaction. "A significant interaction effect" implies that there is variability among cell means over and above that due to variability in row means and column means. Intuitively, we feel there are differences between cell means that cannot be attributed to the collective difference between groups of means from different columns and between groups of means from different rows.

The means from the cells of Table 4.5 are presented in Table 4.9c. Note that the pattern of differences between columns 1 to 4 is different for the three rows; this is most noticeable between row 3 and both of the others. Higher frustration produces higher aggression scores except for those subjects with a low predisposition to aggression (row 3). One can also see from Table 4.9c that the pattern of differences between rows differs from column 1 to 4. This means (1) that the effect (see Section 4.3.1.2) of the experimental-frustration treatment *depends* on the level (row) of aggressive predisposition and (2) that the effect of aggressive predisposition depends on frustration. There is a frustration-treatment effect only for moderately aggressive subjects. Indeed the less aggressive subjects show an order of means that is different from the orders in the other rows. Furthermore, the aggression effect reported significant seems to be much larger for the first two columns than for the last two.

As a consequence of these observations, any discussion of a main effect of aggressiveness would have to be modified—the effect might apply only to the treatment groups for which frustration was greatest. The interpretation of any main effect must take account of any interactions it has with other effects.

The implication for the two-way factorial design is that when there is a significant interaction effect, AB, any statement based on a main effect A is not completely general. That is, either (*a*) there exists some subgroup of subjects corresponding to one or more levels of B for whom the effect is present and others for whom it is absent or (*b*) if the main effect is present for all, it is much greater for some subgroup than it is for some other subgroup.

The interaction in the data of Table 4.5 is shown in graphic form in Figures 4.1a and 4.1b. In Figure 4.1a, values of the aggressiveness factor are shown on the horizontal axis, and values of the dependent variable, *Y*, are on the vertical axis. The four different segmented lines represent the relationship between verbal-aggression score, *Y*, and the aggressiveness variable for each of the four frustration conditions. These graphs should confirm visually the idea that interaction means that different forms of one main effect may occur for various values of the other main effect. In Figure 4.1b, the four frustration conditions are represented on the horizontal axis, and three different segmented lines show the frustration-aggression relationship for the different degrees of aggressiveness. These two figures are alternate ways of representing the two-way interaction.

  4.3.1.4 Summary: Two-way factorial design. When the assumptions for analysis of variance are met, the two-way factorial design (and its analysis) provides the best way to study the effects of two independent variables; it is especially helpful for discovering the modification, if any, on either main effect due to the influence of the other. Such mutual modification, if present, is reflected by the rejection of H_0 for interaction, that is, by a significant interaction.

  The assumptions that must be met for this design are that the scores within cells come from cell populations with normal distributions of scores and with identical variances. If the cell *n*'s are equal

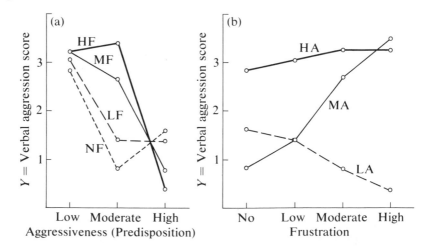

Figure 4.1 (*a*) *Verbal Aggression score as a function of an interaction between the Frustration and Aggressiveness Variables.* (*b*) *Alternative graphic presentation of the interaction.* (*HF = High Frustration; MF = Moderate Frustration; LF = Low Frustration; NF = No Frustration; HA = High Aggression; etc.*)

(as they are in Table 4.5), moderate deviations from these assumptions will not render analysis of variance inappropriate; that is, analysis of variance is "robust against" such moderate deviations. The reader should, however, realize that nearly continuous measurement (see Chapter 1) for Y is needed to insure that deviations from normality are within what we could call a moderate range; a score, Y, that has only three or four possible values cannot have anything like a normal distribution in any population. Despite this fact, if the sample size is large, a variable with a small number of possible values can still be analyzed by means of an analysis of variance model because the deviations from normality will be moderate in this case.

Finally, it should be noted that the term "factorial design" is often used when one of the independent variables is, in fact, not regulated by the experimenter but is, instead, a property of the subject. Indeed, the aggressiveness variable of Table 4.5 is such a variable. For another example, we could study our four frustration treatments for the two sexes producing (what is often called) a two-by-four factorial design. Analysis of variance is appropriate here if the assumptions are still met.

The question to ask is this, "Even though the experimenter cannot assign people independently and at random to sex groups, does his selection of equal numbers of boys and girls do anything systematic to violate the assumptions?" For instance, is it known that the sexes differ on variance or distribution shape for Y, disregarding frustration, the principal independent variable? If so, then the treatments are likely to produce eight cells that suggest gross violation of assumptions. In many instances, variables such as sex, age, social class, educational level, or race can be used without difficulty as factors in factorial designs.

4.3.2 Repeated Measures

When all of the experimental treatments in a study are administered to *all* of the subjects in the study, a two-way design is used; by convention, rows are used for subjects and columns for experimental treatments. For example, we may have each of N subjects learn a list of *paired associates;* this task requires a subject to say a word or syllable out loud when he sees a word or syllable that has been previously paired with it. (If he has seen WUG-TEX, the next time he sees WUG alone, he is to respond with "TEX.") It takes a number of repetitions before he can recall the correct response. For each pair of words or syllables, the number of errors prior to the first correct response can be used as a dependent variable showing rate of learning, as the following example will show.

Suppose we construct a list of nine paired associates. This list contains three sublists that are the experimental treatments. In treatment 1, the paired associates are three pairs of familiar three-letter words (TAG-PUT, BIG-SET, MAT-DOG). In treatment 2, the three paired associates are syllables of the form consonant-vowel-consonant (WIP-CET, MUX-DAK, RIL-HOF). In treatment 3, the three-paired associates are "trigrams" of three consonants (TKQ-LBN, CPJ-RDT, PMC-DHP). The three types of pairs are distributed randomly throughout the list shown to the subject.

Suppose the investigator's theory suggests that the pairs in treatment 3 are most difficult to memorize, those of treatment 1 easiest, and those of treatment 2 intermediate in difficulty. He predicts that he will be able to reject the hypothesis, H_0: $\mu_1 = \mu_2 = \mu_3$. He also predicts H_1: $\mu_3 > \mu_2 > \mu_1$. H_1 shows μ_3 greatest because the greatest number of errors are expected for the trigrams, the next most for the syllables, and the fewest for the words. The problem is how shall he test H_0? His experimental arrangements are presented in Table 4.10, along with hypothetical data. There are 30 subjects but 90 scores, since each subject provides 3 scores, one per treatment. A score, Y_{ij}, represents the number of errors made by subjects i in learning the three paired associates in treatment j. Thus i can be any number from 1 through 30, and j can be 1, 2, or 3.

If H_0 for treatments is true, then both successes and errors will tend to be distributed equally among treatments. The number of errors for the entire list will be apportioned equally among the three kinds of pairs. If the investigator's theory is true, the treatment 1 pairs will be learned early; the Y scores in treatment 1 will tend to be lower than the Y scores for treatment 2; the latter will in turn be lower than the Y scores for treatment 3. The theory predicts $H_1 = \mu_1 < \mu_2 < \mu_3$, which is only one of the many ways in which H_0 can be false. Thus a rejection of H_0 would have to be followed by a test of multiple comparisons (see Section 4.5) to examine the theory's validity.

In any case, it is apparent from Table 4.10 that there is only one score per cell, since a cell is defined by a combination of subject and treatment. Thus there will be no "within-cells" variation. The total variance, S^2_{total} is thus identical to what is called "between-cells" variance, S^2_{between}, in the factorial design. The sum of squares for cells is divided algebraically into parts due to sources for individual subjects (rows), for treatments (columns), and for subject $\times$ treatment interaction. The variance estimate—or mean square—for interaction

Table 4.10 *Subjects by treatments design for the study of performance on a paired-associates task: trials-to-criterion scores*

Subject	Treatment			
	Words	*Syllables*	*Trigrams*	*Total score*
1	4	6	10	20
2	2	7	6	15
3	4	5	7	16
4	5	4	5	14
5	6	7	8	21
6	7	6	7	20
7	4	5	6	15
8	3	6	9	18
9	4	8	8	20
10	4	9	9	22
11	5	6	6	17
12	7	5	5	17
13	8	4	6	18
14	6	7	7	20
15	5	6	8	19
16	4	7	9	20
17	5	8	6	19
18	4	6	8	18
19	4	5	6	15
20	6	4	5	15
21	6	3	6	15
22	4	5	8	17
23	3	4	7	14
24	4	5	7	16
25	3	8	9	20
26	5	7	8	20
27	4	5	9	18
28	6	4	6	16
29	4	5	9	18
30	5	7	6	18

is used as the error term in the denominator for an F test of H_0 for treatments. This reason for this is technical; it is related to the fact that the pattern of "easy versus difficult" tasks will vary among subjects. This means that the mean squares for treatments as well

Table 4.11 *Analysis of variance for three subjects from Table 4.10*

Subject number in Table 4.10	Words	Syllables	Trigrams	Total
12	7	5	5	17
13	8	4	6	18
14	6	7	7	20
$\sum_{i=1}^{3} Y_{ij}$	21	16	18	$\sum_{i=1}^{3}\sum_{j=1}^{3} Y_{ij} = 55$
$\sum_{i=1}^{3} Y_{ij}^2$	149	90	110	$\sum_{i=1}^{3}\sum_{j=1}^{3} Y_{ij}^2 = 349$
$\bar{Y}_j$	7.00	5.33	6.00	
S_j	.817	1.527	.817	

Calculations of sums of squares

$$SS_{total} = \sum_{i=1}^{3}\sum_{i=1}^{3} Y_{ij}^2 - \left(\sum_{i=1}^{3}\sum_{i=1}^{3} Y_{ij}\right)^2 / 3N = 349 - \tfrac{1}{9}(55)^2 = 12.667$$

$$SS_{between\;subjects} = \frac{1}{n}\sum_{j=1}^{3}\left(\sum_{j=1}^{3} Y_{ij}\right)^2 - \frac{1}{Nn}\left(\sum_{i=1}^{3}\sum_{j=1}^{3} Y_{ij}\right)^2$$

$$= \tfrac{1}{3}[(17)^2 + (18)^2 + (20)^2] - \tfrac{1}{9}(55)^2$$

$$= \tfrac{1}{3}[289 + 324 + 400] - 336.333$$

$$= \tfrac{1}{3}(1013) - 336.333$$

$$= 337.667 - 336.333 = \mathbf{1.334}$$

$$SS_{between\;treatments} = \frac{1}{N}\sum_{j=1}^{3}\left(\sum_{i=1}^{3} Y_{ij}\right)^2 - \frac{1}{nN}\left(\sum_{i}\sum_{j} Y_{ij}\right)^2$$

$$= \tfrac{1}{3}[(21)^2 + (16)^2 + (18)^2] - \tfrac{1}{9}(55)^2$$

$$= \tfrac{1}{3}[441 + 256 + 324] - 336.333$$

$$= \tfrac{1}{3}(1021) - 336.333$$

$$= 340.333 - 336.333 = \mathbf{4.000}$$

By subtraction,

$$SS_{int.} = SS_{total} - SS_{treatments} - SS_{subjects}$$
$$= 12.667 - 1.334 - 4.000 = 7.333$$

Summary of analysis of variance

Source	df	SS	MS	f
Treatments	2	4.000	2.000	1.091*
Subjects	2	1.334	.667	<1*
Interaction	4	7.333	1.833	
TOTAL	8	12.667		

*Not significant

Table 4.12   *Summary of the analysis of variance of Table 4.10*

Source	df	SS	MS	f
Subjects	29	48.10	1.659	<1
Treatments	2	94.20	47.100	21.715*
Subject × treatment interaction (error)	58	125.80	2.169	
TOTAL	89	258.10		

*Significant at $\alpha < .01$.

as for interaction contain variation due to idiosyncratic responses of individual subjects to the task. Their ratio, F, will therefore exceed 1.00 just in case the mean square for treatments also contains sizeable variation over and above interaction variance.

To see the influence of these idiosyncratic factors in concrete form, compare subjects 12, 13, and 14 in Table 4.10: note that their patterns of response to the three treatments differ and that for them alone there appear to be no real treatments effect. This is verified by a miniature analysis of variance in Table 4.11.

The summary of analysis of variance for Table 4.10 is presented in Table 4.12. The means and standard deviations of the three treatments are given in Table 4.13. Notice that H_0 for treatments was rejected. From Table 4.13, it would appear that the investigator's prediction of $\mu_1 < \mu_2 < \mu_3$ is correct; the hypothesis based on this may be tested by a multiple-comparisons procedure.

It might further be observed that the row effect was not tested and that it does not appear in any obvious way in Table 4.13. This is because the fact that individuals differ over and above their individual reactions to the tasks (and their total scores differ as well as their patterns) has little bearing on analysis of variance. This fact might be of great interest to a researcher studying the correlates of aptitude for learning paired associates; however, it is not of immediate interest to our hypothetical investigator, who is studying general laws relating task properties and ease of learning.

Table 4.13   *Treatment means and standard deviations for the example of Table 4.10*

Statistic	Treatments		
	Words	Syllables	Trigrams
Mean	4.70	5.80	7.20
SD	1.32	1.45	1.40

4.3.3 A Note on Error

Individual differences constitute error variance in fixed-effects designs and part of the individual differences (the interaction with treatments) are error variance in the repeated-measures design.

It might be well to note that in a different study, say with the temporal variable, rate of presentation, used for the treatments, the treatments mean square might be made up in large part of what

Table 4.14a   *A three-by-three factorial study of the effect of Table 4.10's treatments and rate-of-presentation on errors*

Rate	Treatments		
	Words	*Syllables*	*Trigrams*
Two seconds	6	8	9
	4	6	7
Four seconds	5	6	8
	4	5	7
Six seconds	2	4	6
	2	5	7

Table 4.14b   *Cell means and standard deviations for Table 4.14a.*

Rate	*Words*	*Syllables*	*Trigrams*	Row means
Two seconds	Mean 5.00 SD 1.00	7.00 1.00	8.00 1.00	6.67
Four seconds	Mean 4.50 SD 0.500	5.50 0.500	7.50 0.500	5.83
Six seconds	Mean 2.00 SD 0.00	4.50 0.500	6.50 0.500	4.33
Col. means	3.83	5.67	7.33	$5.61 = \bar{Y}$

Table 4.14c   *Summary of analysis of variance for data of Table 4.14a.*

Source	df	SS	MS	f
Treatments	2	36.778	18.389	19.364*
Rate	2	16.778	8.389	8.877*
Interaction	4	2.222	.556	<1.000†
Within cells	9	8.500	.945	
Total	17	64.278		

*$\alpha < .01$.   †Not significant.

in our example was error. (One person's error is another person's treatment. This is true for all of the designs discussed in this chapter—and generally, for all analyses of variance.)

To clarify this point, a miniature analysis of variance using a fixed-effects factorial study of Table 4.10's treatments and of rate of presentation is presented in Tables 4.14a, b, and c. In this example, there are 18 subjects, each providing one score. Here there is a significant difficulty effect ($F = 19.36; p < .01$); there is no significant interaction, but the rate effect is also significant ($F = 8.88, p < .05$).

In Table 4.15a, I have recast the data as a one-way fixed-effects analysis with only one independent variable, Rate. In the computations and analysis shown in Table 4.15c, the mean square for "within groups" contains the terms that previously contributed (Table 4.14) not only to the "within-cells" mean square but also to the mean square for difficulty and the mean square for the interaction between the main effects. Thus, the F for a test of H_0 in Table 4.15c has as its

Table 4.15a *A one-way analysis of variance with rate as the independent variable (derived from Table 4.14a in which Y = errors)*

Two seconds	Four seconds	Six seconds
6	5	2
4	4	2
8	6	4
6	5	5
9	8	6
7	7	7

Sums

$\sum_{i=1}^{6} Y_{ik}$ 40	35	26 $\sum_{k=1}^{3} \sum_{i=1}^{n} Y_{ik} = 101$

Table 4.15b *Means and standard deviations for Table 4.15a.*

$\bar{Y}_k$ 6.67	5.83	4.33 $\bar{Y} = 5.61$
SD 1.61	1.34	1.89

Table 4.15c *Summary of analysis of variance for Table 4.15a.*

Source	df	SS	MS	F
Rate	2	16.778	8.389	2.572*
Within groups	15	47.500	3.167	
TOTAL	17	64.278		

*$\alpha > .05$; not significant.

denominator (error) term, a mean square made up of three parts. Two of these parts would have been numerator terms in the F's for the Table 4.14 analysis. The df for error is also the total of three df's from Table 4.14c, so the MS would not be larger than before unless the rate and interaction variances were substantial.

"Error" is the name we give to mean squares that estimate that part of variability between subjects that has not been systematically correlated to ("confounded with") the independent variables chosen by us for study. "Error" is thus simply unexplained variability among subjects. In the analysis in Table 4.10, error was composed in part of variance due to Rate, an independent variable in the alternative analysis of Table 4.15.

4.4 More Complex Analyses

4.4.1 Types of Effects

All one-way analyses of variance may be classified into (a) fixed-effects analyses and (b) random-effects analyses. To review, a fixed-effects variable is one whose categories or levels are designed by the experimenter; these categories constitute, conceptually, all there are. A random-effects variable is one for which the categories used in the experiment represent a random sample from some larger number of categories. In my discussion so far, the only random effects illustrated have been those for between subjects in the repeated-measures design (Section 4.3.2). I treated that as a *two*-way design with subjects constituting one variable, which we now see is a random effect. Examples of other random effects are effects of experimenters, judges, raters, scorers, test items, and litters of animals.

Complex designs (having more than one independent variable) may be constructed from combinations of two or more variables. Any design, two-, three-, four-, or more-way, in which all variables are fixed effects is called a *factorial design* and treated in a manner similar to the two-way factorial design. An example of a four-way factorial design and the summary of the analysis appears in Tables 4.16a, 4.16b, and 4.16c.

For this example, the variables are Motivational Set (A), Story (B), Hero Sex (C), Child Sex (D). There are three children per combination, and the dependent variable is a rating of emotionality. Note that the sums of squares are calculated by squaring the total of the scores in a unit, dividing by the number of scores in the unit, and then subtracting a constant equal to the grand total T, squared and divided by N. This pattern is repeated from effect to effect.

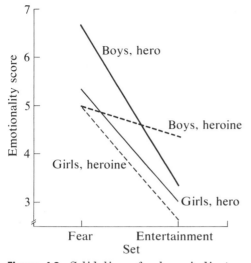

Figure 4.2 *Solid lines for boys indicate a marked two-way interaction between set and hero sex. Emotionality is highest for the same-sex hero (male) only under fear set. Girls uniformly have slightly more emotionality for male heros regardless of set. Thus,* the two-way interaction between set and hero sex is present for boys but absent for girls. *This is what a significant three-way interaction between set, hero sex, and child sex means.*

The only significant interaction is depicted in Figure 4.2. Interactions involving more than two variables are difficult to conceptualize, and Figures such as Figure 4.2 are meant to assist in this task.

Complex interactions are often hard to conceptualize and to use for any practical purpose. Some investigators feel that very complex interactions amount to "error" individual differences because the number of combinations of variables required to explain them is so large that little insight is gained from the explanation. On the basis of this logic, it is sometimes suggested that the sums of squares for complex interactions be added to the sum of squares for error. In the data of Table 4.16a, this would mean adding sums of squares for $A \times B \times C \times D$ to the "within" sum of squares. This pooled SS could then be divided by the total df for error and $A \times B \times C \times D$ interaction. The resulting F for a simple effect, such as "sex of subject," would be referred to the F table for test.

Table 4.16a *A two × two × two × two factorial design with Y = "emotionality index"*

Main character	Fear Set		Entertainment Set		$\sum Y$
	Story 1	Story 2	Story 1	Story 2	

Boys

Hero	7 6 7	8 5 7	4 3 3	4 3 4	
	$\sum Y = 20$	$\sum Y = 20$	$\sum Y = 10$	$\sum Y = 11$	
	$\sum Y = 40$		$\sum Y = 21$		$\sum Y = 61$
Heroine	6 5 4	6 5 4	4 5 4	4 4 3	
	$\sum Y = 15$	$\sum Y = 15$	$\sum Y = 13$	$\sum Y = 11$	
	$\sum Y = 30$		$\sum Y = 24$		$\sum Y = 54$
	$\sum Y = 70$		$\sum Y = 45$		
		115			

Girls

Hero	6 6 4	6 5 4	2 4 3	2 4 3	
	$\sum Y = 16$	$\sum Y = 15$	$\sum Y = 9$	$\sum Y = 9$	
	$\sum Y = 31$		$\sum Y = 18$		$\sum Y = 49$
Heroine	6 4 5	5 4 6	2 3 3	2 3 3	
	$\sum Y = 15$	$\sum Y = 15$	$\sum Y = 7$	$\sum Y = 7$	
	$\sum Y = 30$		$\sum Y = 14$		$\sum Y = 44$
	$\sum Y = 61$		$\sum Y = 32$		
		93			
	$\sum Y = 131$		$\sum Y = 77$		$\sum Y = 208$

$$T = 208$$

$N = 48$ $\qquad$ $T^2 = 43264$

$\text{Total} \sum Y^2 = 950$ $\qquad$ $G = T^2/N = \mathbf{901.333}$

Calculations

$SS_{\text{total}} = 1010 - 901.333 = \mathbf{108.667}$

$SS_{ABCD \text{ combs.}} = \frac{1}{3}\left(\frac{984}{2952}\right) - 901.333 = 984 - 901.333 = \mathbf{82.667}$
(cells)

(Continued)

Table 4.16a (Continued)

Calculations (Continued)

$SS_{ABC\,comb.} = \frac{1}{6}[(36)^2 + (35)^2 + (19)^2 + (20)^2 + (30)^2 + (30)^2 + (20)^2 + (18)^2] - G$
$\frac{1}{6}(5806) - 901.333 = 967.667 - 901.333 = \mathbf{66.334}$

$SS_{ABD\,comb.} = \frac{1}{6}[(35)^2 + (35)^2 + (23)^2 + (22)^2 + (31)^2 + (30)^2 + (16)^2 + (16)^2] - G$
$\frac{1}{6}(5836) - 901.333 = 972.667 - 901.333 = \mathbf{71.334}$

$SS_{BCD\,comb.} = \frac{1}{6}[(30)^2 + (31)^2 + (28)^2 + (26)^2 + (25)^2 + (24)^2 + (22)^2 + (22)^2] - G$
$\frac{1}{6}(5490) - 901.333 = 915.000 - 901.333 = \mathbf{13.667}$

$SS_{ACD\,comb.} = \frac{1}{6}[(40)^2 + (21)^2 + (30)^2 + (24)^2 + (31)^2 + (18)^2 + (30)^2 + (14)^2] - G$
$\frac{1}{6}(5898) - 901.333 = 983 - 901.333 = \mathbf{81.667}$

$SS_{AB\,comb.} = \frac{1}{12}[(66)^2 + (65)^2 + (39)^2 + (38)^2] - G$
$\frac{1}{12}(11546) - 901.333 = 962.167 - 901.333 = \mathbf{60.834}$

$SS_{AC\,comb.} = \frac{1}{12}[(70)^2 + (60)^2 + (39)^2 + (38)^2] - G$
$\frac{1}{12}(11606) - 901.333 = 967.167 - 901.333 = \mathbf{65.834}$

$SS_{AD\,comb.} = \frac{1}{12}[(70)^2 + (45)^2 + (61)^2 + (32)^2] - G$
$\frac{1}{12}(11670) - 901.333 = 972.500 - 901.333 = \mathbf{71.167}$

$SS_{BC\,comb.} = \frac{1}{12}[(55)^2 + (55)^2 + (50)^2 + (48)^2] - G$
$\frac{1}{12}(10854) - 901.333 = 904.500 - 901.333 = \mathbf{3.167}$

$SS_{BD\,comb.} = \frac{1}{12}[(58)^2 + (57)^2 + (47)^2 + (46)^2] - G$
$\frac{1}{12}(10938) - 901.333 = 911.500 - 901.333 = \mathbf{10.167}$

$SS_{CD\,comb.} = \frac{1}{12}[(61)^2 + (54)^2 + (49)^2 + (44)^2] - G$
$\frac{1}{12}(10974) - 901.333 = 914.500 - 901.333 = \mathbf{13.167}$

$SS_A = \frac{1}{24}[(131)^2 + (77)^2] - G$
$\frac{1}{24}(23090) - 901.333 = 962.083 - 901.333 = \mathbf{60.750}$

$SS_B = \frac{1}{24}[(105)^2 + (103)^2] - G$
$\frac{1}{24}(21634) - 901.333 = 901.417 - 901.333 = \mathbf{0.084}$

$SS_C = \frac{1}{24}[(110)^2 + (98)^2] - G$
$\frac{1}{24}(21704) - 901.333 = 904.333 - 901.333 = \mathbf{3.000}$

$SS_D = \frac{1}{24}[(115)^2 + (93)^2] - G$
$\frac{1}{24}(21874) - 901.333 = 911.417 - 901.333 = \mathbf{10.084}$

$SS_{AB} = SS_{AB\,comb.} - SS_A - SS_B = 60.834 - 60.750 - 0.084 - \mathbf{0.000}$

$SS_{AC} = SS_{AC\,comb.} - SS_A - SS_C = 65.834 - 60.750 - 3.000 = \mathbf{2.084}$

$SS_{AD} = SS_{AD\,comb.} - SS_A - SS_D = 71.167 - 60.750 - 10.084 = \mathbf{0.333}$

$SS_{BC} = SS_{BC\,comb.} - SS_B - SS_C = 3.167 - 0.084 - 3.000 = \mathbf{0.083}$

$SS_{BD} = SS_{BD\,comb.} - SS_B - SS_D = 10.167 - 0.084 - 10.084 = \mathbf{0.000}$ (rounded)

$SS_{CD} = SS_{CD\,comb.} - SS_C - SS_D = 13.167 - 3.000 - 10.084 = \mathbf{0.083}$

$SS_{ABC} = SS_{ABC\,comb.} - SS_{AB} - SS_{BC} - SS_{AC} - SS_A - SS_B - SS_C$
$= 66.334 - 0.000 - 0.083 - 2.084 - 60.750 - 0.084 - 3.000 = \mathbf{0.333}$

$SS_{ABD} = SS_{ABD\,comb.} = SS_{AB} - SS_{BD} - SS_{AD} - SS_A - SS_B - SS_D$
$= 71.334 - 0.000 - 0.000 - 0.333 - 60.750 - 0.084 - 10.084 = \mathbf{0.083}$

$SS_{BCD} = SS_{BCD\,comb.} - SS_{BC} - SS_{CD} - SS_{BD} - SS_B - SS_C - SS_D$
$= 13.667 - 0.083 - 0.083 - 0.000 - 0.084 - 3.000 - 10.084 = \mathbf{0.333}$

$SS_{ACD} = SS_{ACD\,comb.} - SS_{AC} - SS_{CD} - SS_{AD} - SS_A - SS_C - SS_D$
$= 81.667 - 2.084 - 0.083 - 0.333 - 60.750 - 3.000 - 10.084 = \mathbf{5.333}$

$SS_{ABCD} = SS_{ABCD\,comb.} - SS_{ABC} - SS_{ABD} - SS_{BCD} - SS_{ACD} - SS_{AB} - SS_{AC}$
$- SS_{AD} - SS_{BC} - SS_{BD} - SS_{CD} - SS_A - SS_B - SS_C - SS_D = \mathbf{0.084}$

$SS_{within} = \sum SS_{w/ijkl} = 26.003$

$SS_{within\,check} = SS_{total} - SS_{cells} = 108.667 - 82.667 = 26.000$ (checks to two places)

Table 4.16b *Means and standard deviations of cells*

Main character	Fear		Entertainment	
	Story 1	Story 2	Story 1	Story 2
	Boys			
Hero	Mean 6.67	6.67	3.33	3.67
	SD .472	1.245	.472	.472
Heroine	Mean 5.00	5.00	4.33	3.67
	SD .817	.817	.472	.472
	Girls			
Hero	Mean 5.33	5.00	3.00	3.00
	SD .943	.817	.817	.817
Heroine	Mean 5.00	5.00	2.67	2.67
	SD .817	.817	.472	.472

Main effect means ($\bar{Y}_1$ = mean of level 1 for the effect, i.e., the condition at left or at the top in the main table)

Effect	$\bar{Y}_1$	$\bar{Y}_2$
Set	5.46	3.21
Story	4.38	4.29
Hero sex	4.58	4.08
Child sex	4.79	3.88

Table 4.16c *Summary of analysis of variance*

Source	df	SS	MS	F^*
Set (A)	1	60.750	60.750	74.723
Story (B)	1	0.084	0.084	
Hero(ine) sex (C)	1	3.000	3.000	
Child's sex (D)	1	10.084	10.084	12.328
$A \times B$	1	0.000	0.000	
$A \times C$	1	2.084	2.084	
$A \times D$	1	0.333	0.333	
$B \times C$	1	0.083	0.083	
$B \times D$	1	0.000	0.000	
$C \times D$	1	0.083	0.083	
$A \times B \times C$	1	0.333	0.333	
$A \times B \times D$	1	0.083	0.083	
$A \times C \times D$	1	5.333	5.333	6.560
$B \times C \times D$	1	0.333	0.333	
$A \times B \times C \times D$	1	0.084	0.084	
Within cells (error)	32	26.003	0.813	
TOTAL	47	108.667		

*Only those F's significant at $\alpha < .05$ are presented.

In this example, the new pooled-error term is .791, and the F for sex (12.736) is about the same size as the old one (12.328); however, the denominator df has been increased from 32 to 33, making it slightly easier for an F of fixed size to reach a preset significance level. This pooling procedure is based on the assumption that none of the sums of squares added to the "within" term represent real interactions—that is, one must assume that they represent individual differences *of the same kind* as "error." This assumption is precarious. Whenever it is not true, the probability of Type II error is always *increased*.

In some complex designs, one or more of the effects to be tested is a random effect. Such designs are designated as "mixed." In Tables 4.17a, 4.17b, and 4.17c, the data and analysis for a mixed three-way design are presented. The random effects are Subjects and the fixed effects are Set and Task. Thus each subject presumably performs each of two tasks under each of the two motivational sets, Achievement and Speed. In such a situation, steps are often taken to see that half of the subjects perform first under one set and next under the other set; the other subjects follow the reverse order. This procedure is called counterbalancing the order of treatments and is a desirable one to follow. However, there is no way to tell from a table like Table 4.17a whether counterbalancing was done or not. The dependent variable is error. Note that although the arithmetic is similar to that described for the complex factorial design, the analysis (Table 4.17c) is different; there is no "within" term and error individual differences are represented by the interactions of the effect to be tested with subjects. Thus, Task effect is tested by an F with the Task $\times$ Subject interaction in the denominator; Set is tested by an $F = MS_{\text{set}}/MS_{\text{set} \times \text{subj.}}$; Set $\times$ Task interaction is tested by $F = MS_{\text{set} \times \text{task}}/MS_{\text{set} \times \text{task} \times \text{subj.}}$. The two-way interactions of set and task with subjects are not tested, and their use as error terms for tests of main effects represents *an assumption that these terms represent only unexplained or error individual differences and that there is no meaningful interaction between subjects and fixed effects.* Where it is feasible to administer all treatment combinations to all subjects, this design is preferred because the (random) Subjects effect can be parceled out of the error terms. As a consequence of this fact, the error terms contain only those individual differences that are related to the fixed effect being tested. This preferred procedure is an example of a repeated-measurements experiment, so named because more than one measurement is taken from each subject.

There are many other complex designs besides the two illustrated here. The arithmetic is identical for all k-way designs. Tests of different hypotheses by means of the F-distribution require the

Table 4.17a   *Mixed three-way design for analysis of variance performance scores*

Subject	Achievement		Speed		Totals for subjects
	Task 1	Task 2	Task 1	Task 2	
1	12	10	12	10	44
2	10	9	13	11	43
3	9	8	15	14	46
4	10	9	10	9	38
5	8	6	11	11	36
6	6	6	9	10	31
7	5	4	8	9	26
8	7	6	12	11	36
9	8	5	13	12	38
10	10	8	15	14	47

Sums and Means

$\sum Y$	85	71	118	111	$\sum Y = 385$
$\sum Y^2$	763	539	1442	1261	$\sum Y^2 = 4005$
$\bar{Y}_{jk}$	8.5	7.1	11.8	11.1	$\bar{Y} = 9.63$
					$JKN = 40$

Calculations

$G = T^2/JKN = (385)^2/40 = 3705.625.$

$SS_{\text{total}} = 4005 - 3705.625 = 299.375$

$SS_{\text{between }S's} = \frac{1}{4}[(44)^2 + (43)^2 + (46)^2 + (38)^2 + (36)^2 + (31)^2 + (26)^2 + (36)^2 + (38)^2 + (47)^2] - G$
$\qquad = \frac{1}{4}(15227) - G = 3806.750 - G = \mathbf{101.125}$

$SS_{\text{task}} = \frac{1}{20}[(203)^2 + (182)^2] - G = \frac{1}{20}(74333) - G = 3716.667 - G = \mathbf{11.042}$

$SS_{\text{set}} = \frac{1}{20}[(156)^2 + (229)^2] - G = \frac{1}{20}(76777) - G = 3838.850 - G = \mathbf{133.225}$

$SS_{\text{set}\times S \text{ combs.}} = \frac{1}{2}[(22)^2 + (19)^2 + (17)^2 + (19)^2 + (14)^2 + (12)^2 + (9)^2 + (13)^2 + (13)^2 + (18)^2 + (22)^2 + (24)^2 + (29)^2 + (19)^2 + (22)^2 + (19)^2 + (17)^2 + (23)^2 + (25)^2 + (29)^2] - G = \frac{1}{2}(7969) - G = 3984.500 - G = \mathbf{278.875}$

$SS_{\text{task}\times S \text{ comb.}} = \frac{1}{2}[(24)^2 + (23)^2 + (24)^2 + (20)^2 + (19)^2 + (15)^2 + (13)^2 + (19)^2 + (21)^2 + (25)^2 + (20)^2 + (20)^2 + (22)^2 + (18)^2 + (17)^2 + (16)^2 + (13)^2 + (17)^2 + (17)^2 + (22)^2] - G = \frac{1}{2}(7647) - G = 3823.500 - G = \mathbf{117.875}$

$SS_{\text{task}\times\text{set comb.}} = \frac{1}{10}[(85)^2 + (71)^2 + (118)^2 + (111)^2] - G = \frac{1}{10}(38511) - G = 3851.100 - G = \mathbf{145.475}$

$SS_{\text{task}\times S} = SS_{\text{task}\times S \text{ comb.}} - SS_{\text{task}} - S_{\text{set }S's} = 117.875 - 11.042 - 101.125 = \mathbf{5.708}$

$SS_{\text{set}\times S} = SS_{\text{set}\times S \text{ comb.}} - SS_{\text{set}} - SS_{\text{bet. }S's} = 278.875 - 133.225 - 101.125 = \mathbf{44.525}$

$SS_{\text{task}\times\text{set}} = SS_{\text{task}\times\text{set comb.}} - SS_{\text{task}} - SS_{\text{set}} = 145.475 - 11.042 - 133.225 = \mathbf{1.208}$

$SS_{\text{task}\times\text{set}\times S} = SS_{\text{total}} - SS_{\text{task}\times\text{set}} - SS_{\text{task}\times S} - SS_{\text{set}\times S} - SS_{\text{set}} - SS_{\text{task}} - SS_{\text{bet. }S's}$
$\qquad = \mathbf{2.542}$

Table 4.17b *Means and standard deviations for Table 4.17a*

Statistic	Achievement set		Speed set	
	Task 1	Task 2	Task 1	Task 2
Mean	8.5	7.1	11.8	11.1
SD	2.01	1.87	2.22	1.70

Table 4.17c *Summary of the analysis of variance of Table 4.17a*

Source	df	SS	MS	f*
Set (A)	1	133.225	133.225	26.930
Task (B)	1	11.042	11.042	17.416
Subjects (S)	9	101.125	11.236	
A × B	1	1.208	1.208	4.284
A × S	9	44.525	4.947	
B × S	9	5.708	0.634	
A × B × S	9	2.542	0.282	
TOTAL	39	299.375		

* All f's presented significant at α less than .05. Interactions with S are treated as error variance and customarily are not tested. This latter point is discussed in the text.

assumption that the simplest *units*—cells, columns, etc.—in the error term represent subpopulations with equal variances and normally distributed scores. They also require that the expected values of numerator and denominator terms in an F differ *only* by the variance due to the tested effect. This last assumption sometimes requires that true variances for some interaction(s) be assumed to be equal to zero.

4.5 Scheffe Multiple-Comparisons Procedures following Complex Analysis of Variance

The procedure discussed in Section 4.2 for comparing individual means following a one-way analysis of variance can be extended to two-, three-, and K-way analyses quite readily. Any two means (or any two properly weighted combinations of means) involved in the *same effect* can be calculated and then divided by the square root of a function of the *MS used as error* in the *F* test of that effect. Thus, from Table 4.17a, we can select for comparison Tasks 1 and 2 only under the Achievement set. The critical ratio we need is

$$\frac{\bar{Y}_{\text{task 1, ach.}} - \bar{Y}_{\text{task 2, ach.}}}{\sqrt{MS_{\text{task} \times \text{subj.}}\left(\frac{1}{10} + \frac{1}{10}\right)}}$$

It is compared to the critical value (*C.V.*)

$$C.V. = \sqrt{df_{\text{task}}F_{\alpha}}$$

which is analogous to the value given in Section 4.2 for the Scheffe procedure following a one-way analysis. The critical value for $\alpha = 0.05$ is computed as $\sqrt{1 \times 4.08} = 2.02$, and the critical ratio (*C.R.*) as

$$C.R. = \frac{8.5 - 7.1}{\sqrt{0.634\left(\frac{1}{10} + \frac{1}{10}\right)}} = \frac{1.400}{\sqrt{0.1268}} = \frac{1.400}{0.357} = 3.92$$

from which we see that the latter is larger than the former; thus we reject the hypothesis of no difference between the tasks under the achievement set.

4.6 A Note on Analysis of Covariance

An investigator studying one independent variable, X, may wish to control the influence of some second variable, Z, which is known to be correlated with the dependent measure, Y, but which cannot be used as a second independent variable; this is the case when it is not possible to assign subjects at random to values of Z, as when Z is a measured property of the subject.

For example, suppose we are studying the effects of four different types of preschool experience on IQ test score after one year of exposure. In such a study (*a*) the initial IQ score before the preschool experience would influence final outcome; (*b*) the four treatments might work differently for different intelligence levels; (*c*) we obviously cannot assign children to initial IQ scores; (*d*) it may well be that the treatment groups cannot be constituted so that they have similar means, standard deviations, and distribution shapes for IQ score. (If we use pre-existing classes, the latter problem is almost always present; even with random assignment from a limited subject pool, it is likely to be present.)

In this study, it would be appropriate to analyze the variance of *that portion of final IQ that cannot be predicted from initial IQ*. One procedure for doing this is called "covariance analysis" and consists in subtracting from each individual score, Y_i (final IQ), that portion of it, Y_i', that is predictable from Z_i (initial IQ) and then computing the usual analysis of variance on the resulting $(Y - Y')$'s. (Adjustments must be made to the degrees of freedom due to the fact that estimation of the regression coefficient, B_{yz}, required loss of degrees of freedom. Otherwise, the procedure is simply an analysis of variance for the adjusted scores $(Y_i - Y_i')$'s, with the independent variable, X, being preschool experience.)

The use of covariance requires one to assume that a given type of relationship (for example, linear) holds between the dependent variable and Z. Furthermore, one assumes that this form of relationship is the same in the various treatment groups. Analysis of covariance is especially appropriate in situations in which *change scores* (posttest minus pretest) might otherwise be considered. Such change scores often fail to meet the basic assumptions of analysis of variance–normality and equal variance across treatments. On the other hand, pretest and posttest scores often do relate in a linear way within treatment groups.

4.7 Summary: Analysis of Variance

Analysis of variance is basically an arithmetical technique for dividing the estimate of total variation among individuals into parts that are meaningful in the context of the investigator's procedures. Hence, we have in a summary table a *list* of various sources of variation together with estimates (mean squares) of the amounts of the total variance attributable to them. These sources are the various combinations of treatments or of treatments and other classifications, such as age and sex. Insofar as we have chosen these treatments wisely, the resultant list and accompanying tests of significance will be helpful in providing insight into the nature of the dependent variable.

An analysis of variance of a dependent variable such as aggressiveness will be appropriate only if the investigator chooses experimental treatments and control variables—sources—that are indeed systematically related to Y. This choice requires good judgment; it depends on factors unrelated to the arithmetic of analysis of variance. It may depend on how well the investigator has done his homework in the field of aggression; it may also depend on the investigator's interests, on his creative hunches, and on leads from the work of others. Your tasks are, *first,* to understand the investigator's question(s) and, *second,* to use the results of analysis of variance to evaluate the investigator's answer(s).

In this chapter, I have tried to provide the guidelines necessary to these tasks. Insofar as your questions are different from those of the investigator, a given analysis may be inadequate to your purpose. A second use, then, of skill in interpreting analysis of variance results is the ability to compare and relate studies that ask different questions. You can use your skill to relate (or contrast) one investigator's sources to another's sources as you develop your own interpretation of a field of interest.

5

Multivariate Analyses

5.1 Introduction

It is helpful, in the analysis of many problems, to have a number of scores for each subject. The classic example occurs in the field of intelligence testing: a long-held theory says that a general intelligence, g, is reflected in a variety of specific performance measures. To study intelligence in the context of that theory, it is natural to administer many tests of mental skills, such as vocabulary, speed of recall, verbal analogies, mental arithmetic, space perception. The score on each test is one variable, X_i, and there are several, k, of such scores for each subject, represented as $X_1, X_2, X_3, X_4 \ldots X_k$.

Techniques that take account of, or specifically focus on, the various relationships among variables are termed "multivariate analyses." The simplest to perform is the inspection of a table showing all the possible correlation coefficients between pairs of variables from the total set. This is often done, before additional analyses are made, in order to discover groupings or clusters of variables that may be of special interest. At times, one of the k variables is the focus; the problem is to predict its value for each subject from his scores on the remaining $k - 1$ variables. The research problem here is to find coefficients (or "weights") to associate with each of the remaining $k - 1$ variables in order to maximize the accuracy with which the focal variable can be predicted from them. This problem is solved by means of the technique of *multiple regression*. The resulting multivariate-correlation coefficient is termed a *multiple correlation*.

Factor analysis is a third multivariate technique. It is a procedure for reducing complexity; effective study of the correlations between all possible pairs of a large number of variables is not possible. Factor analysis seeks to discover a very few basic variables, called "factors," which explain almost as much about the numerous relationships as do the two-variable (bivariate) correlations. There are limits on what "explain" means in the previous sentence, but in any case, factor analysis does reduce the complexity of correlational data; it effectively reduces the number of variables to be kept in mind from k to some smaller number p. Thus, for example, in a study in which 15 mental tests were administered to 5000 school children, four factors might be found that account for a large portion of whatever it is that the tests have in common.

A fourth multivariate technique is called for when the researcher wishes to relate two broadly defined but distinct sets of variables. Suppose, for example, that one believed grade school adjustment to be related to the health and physical maturity of the child. For each child, there would be a number of adjustment scores—tests, teacher's ratings, sociometric ratings, parents' ratings and/or reports, and so forth. Similarly, for health and physical maturity, one might have a number of scores such as heart rate, height, weight, basal metabolism, an index of intensity of illnesses, and pediatrician's ratings. *Canonical correlation* is a technique used to discover factors separately in the adjustment set and in the health-and-maturity set such that the multiple correlation between sets of factors will be the maximum possible.

Another problem is that of assigning weights to the variables in sets in such a way as to maximize the mean difference between groups with respect to the scores constructed from (linear) combinations of the variables. When the weights are found, one can predict group membership with maximum success. The technique for doing this is called *discriminant analysis;* it is often applied to practical problems of vocational and academic guidance.

This chapter will be devoted to an elementary discussion of the first three of the five techniques outlined in the above paragraphs and to general issues of interpretation common to all multivariate techniques. The first section introduces the concept "common variance." The next five sections deal with specific techniques and the last one gives critical attention to issues of statistical reliability, validity, evaluation of results, the use of multivariate analysis in theory building, and the problem of arbitrariness in selecting input measures.

A number of specific techniques—multidimensional scaling, multivariate analysis-of-variance, cluster-and-type analysis, laten-structure analysis, moderated prediction (see Section 5.4), nonlinear multiple regression—are not discussed in this book; I see the broad issues and problems raised by the omitted techniques as similar to those raised by the techniques that will be discussed.

5.2 The Concept "Common Variance"

Multivariate problems are often stated in terms of accounting for some variance—the variance of a criterion variable (multiple correlation), or the total of the variances in a set of variables (factor analysis). In reports of multivariate research, the reader is likely to note an emphasis on *common variance*, and, especially, on just how much of this common variance a given analysis was able to account for. In this section, we will deal with the questions, "What is common variance?", "Why is it important?", and "How can one set about accounting for it?"

Let us begin with the simplest case of two variables, X and Y; we shall use V to stand for the total of their variances.[1] Thus,

$$V = S_x^2 + S_y^2 \tag{5-1}$$

A certain portion of V is said to be common to X and Y. The remaining part consists of two portions, one specifically due to X and one specifically due to Y. Thus an analysis of X and Y begins with the notion expressed as

$$V = S_{\text{common}}^2 + S_{x_{\text{specific}}}^2 + S_{y_{\text{specific}}}^2 \tag{5-2}$$

---

[1] V is *not* the same as the variance of $X + Y$. The latter sample variance is given as

$$S_{X+Y}^2 = S_X^2 + S_Y^2 + r_{XY} S_X S_Y$$

$S_{X+Y}^2 = V$ only in case X and Y are uncorrelated, which is almost never the case in multivariate analyses.

It can be shown (see Section 5.4) that

$$S^2_{\text{common}} = r^2_{xy} V \tag{5-3}$$

It might be said that whatever X and Y have in common accounts for (or explains) the common variance, S^2_{common}. For instance, if X were height and Y weight we might venture the opinion that physical size explains $r^2_{xy} \times 100$ percent of the variance *common* to X and Y. In so doing, we have simply given a name to a hypothetical factor that underlies both X and Y.

If we extend the problem to several variables, the same principle holds. That is, we have $X_1, X_2, X_3, X_4 \ldots X_k$ and

$$V = S^2_{x_1} + S^2_{x_2} + S^2_{x_3} + \cdots + S_{x_k} \tag{5-4}$$

If at least two of the variables are correlated, then part of V is common to *at least* two of the variables, and part of it is made up of parts specific to each of the variables separately. The common variance is customarily symbolized H^2, and referred to as "the communality." Specific variances for $X_1, X_2 \ldots X_i \ldots X_k$ are symbolized $S^{*2}_{X_1}, S^{*2}_{X_2} \ldots S^{*2}_X$; these variances represent that portion of X_i variation not related to variation in any other variable. Thus,

$$V = H^2 + S^{*2}_{X_1} + S^{*2}_{X_2} + \cdots + S^{*2}_{X_k} \tag{5-5}$$

Tables 5.1, 5.2, and 5.3 further illustrate the common variance notions. For simplicity, standardized scores are assumed; thus the total variance in each table is 3. Table 5.1 shows the intercorrelations among three variables with little common variance. Note that all three correlations are small and thus that no two of the three variables have high common variance (r^2). Table 5.2 shows the intercorrelations among three variables with moderate common variance. Note here that two of the variables, IQ Score and College Aptitude Test correlate highly and thus have moderately high common variance

Table 5.1   *Intercorrelations among three variables with no common variance*

Variable	IQ score	Height	Musical aptitude score
IQ	1.000	0.032	0.092
Height		1.000	0.128
Musical aptitude			1.000

Table 5.2 *Intercorrelations among three variables with moderate common variance*

Variable	IQ score	Height	College aptitude test
IQ	1.000	0.032	0.902
Height		1.000	.036
College aptitude			1.000

between them. However, neither IQ Score nor College Aptitude Test correlates highly with the third variable, Height; thus the total common variance is modest. In Table 5.3, all three variables, IQ Score, Vocabulary Test, and College Aptitude Test, correlate highly; thus the common variance in all three pairs is high and so is the total common variance. Further indication of how the H^2 figures for these tables would be determined will be presented in Section 5.5.

Multivariate techniques are designed to account for, or to make use of, H^2, the common variance. This is perhaps most apparent in the case of factor analysis, wherein one or more new variables are discovered and defined in such a way as to explain whatever it is the k original variables have in common. In multiple-regression analysis, one often seeks to discover those predictor variables that will maximize the value of H^2; such predictor variables are themselves resources for explaining the variance of the predicted variable. Canonical correlation seeks to explain the common variance in one set of variables by reference to the common variance in another set.

In summary, multivariate techniques are used to discover meaning in one or more members of a set of variables by simultaneous reference to all the other members of the set. To do this is to explain the common variance in the set. The common variance is an arithmetic term representing that portion of the total variance that is due to combinations of two, three, . . . or k of the variables in the set. The

Table 5.3 *Intercorrelations among three variables with high common variance*

Variable	IQ score	Vocabulary test	College aptitude test
IQ	1.000	.851	0.902
Vocab.		1.000	.898
College aptitude			1.000

Table 5.4   *Intercorrelations among eight hypothetical measures of children's personality traits*

Trait	Trait							
	1. Mannerliness	2. Approval Seeking	3. Initiative	4. Guilt	5. Sociability	6. Creativity	7. Adult Role	8. Cooperativeness
1. Mannerliness		.709	.204	.081	.626	.113	.155	.774
2. Approval Seeking			.051	.089	.581	.098	.083	.652
3. Initiative				.671	.123	.689	.582	.072
4. Guilt					.022	.798	.613	.111
5. Sociability						.047	.201	.724
6. Creativity							.801	.120
7. Adult Role								.152
8. Cooperativeness								

common variance can be contrasted with the remaining portion of the total variance; the latter is made up of variation specific to each of the variables separately.

5.3 The Intercorrelation Matrix

The first task for the reader of a multivariate study often is to examine the Pearson product-moment coefficients of correlation between all pairs of variables in the study.[2] These correlations are usually presented in tables such as Table 5.4, 5.6, or 5.8.

5.3.1 A Case in which No r's are Negative

Table 5.4 gives the correlations between pairs of variables in a hypothetical study of teacher's ratings of children's personality characteristics. Let us examine it in detail.

---

[2] Multivariate techniques are sometimes applied to tables of other types of correlation indices. However, the suggestions here given for studying the table of intercorrelations depend on the variance interpretation of r. This interpretation must be qualified or abandoned in the case of some other types of coefficients.

First, we observe a range in magnitude from an r of .801 between Adult Role and Creativity to an r of .022 between Sociability and Guilt. The mean value[3] of the 28 tabled r's is little help to us, since so many different sets of relationships could all be consistent with the same mean. Consequently, it is recommended that after the range is noted, one proceed to pick out specific individual correlations coefficients that will be of most interpretive value. These are those r's that are either quite high or quite low. Our goal here will be to see whether there is any systematic grouping or clustering of the variables suggested by the high-r and low-r lists.

Our first difficulty is readily apparent—How high is "high?" How low is "low?" Admittedly, the decision implied is (1) arbitrary and (2) to be useful, dependent upon the actual range of r values present in a given table. I have found it helpful to use the variance interpretation of r (see Section 5.4). "High" then becomes, at minimum, a value of r that suggests that two variables have a considerable common variance; "low" becomes, at maximum, a value of r that suggests that the variables have virtually no common variance.

I recommend that the interpretation be done in stages. First, jot down all pairs of variables that have at least 50 percent of their total variance in common. These would be all pairs for which $r^2 \geq .500$ or $r \geq .707$. (For Table 5.4, these pairs are shown in List I of Table 5.5.)

Next, jot down all pairs of variables that have less than 5 percent of their total variance in common. These would be those pairs for which $r^2 \leq .05$ or $r \leq .224$. (For Table 5.4, these pairs are shown in List III of Table 5.5.)

Now, examine the "high" list of pairs for possible groupings of three or more variables. Such a grouping tentatively is suggested by Pairs 1 and 2 on the list. Both Adult Role and Guilt are highly related to Creativity. To establish the three variables involved in these two pairs as a set distinct from other variables we move to the "low" list to check for correlations between each of Guilt, Creativity, and Adult Role and any of the remainder. We observe that all three appear with Mannerliness on the "low" list; all three also appear with Approval Seeking; with Cooperativeness; with Sociability.

It would now seem safe to say that Guilt, Adult Role, and Creativity are a group of variables with something in common, and that

---

[3]This mean must be, and has been, determined by performing a transformation on the r's to normalize their sampling distribution. The transformed values are then averaged and the resulting mean is retransformed and expressed as an r. This procedure is necessary if the mean r is to be representative of central tendency of the tabled values.

Table 5.5   *Interpretive lists of pairs of variables from Table 5.4*

I. *Higher correlated pairs: pairs in which more than 50 percent of the total variance is common variance.* $(r \geq .707)$
  1. Adult Role and Creativity $(r = .801)$
  2. Guilt and Creativity $(r = .798)$
  3. Mannerliness and Cooperativeness $(r = .744)$
  4. Sociability and Cooperativeness $(r = .724)$
  5. Mannerliness and Approval Seeking $(r = .709)$

II. *Moderately high correlated pairs: pairs in which more than 25 percent of the total variance, but less than 50 percent of it, is common variance.* $(.500 \leq r \leq .707)$
  1. Initiative and Creativity $(r = .689)$
  2. Initiative and Guilt $(r - .671)$
  3. Approval Seeking and Cooperativeness $(r = .652)$
  4. Mannerliness and Sociability $(r = .626)$
  5. Guilt and Adult Role $(r = .613)$
  6. Initiative and Adult Role $(r = .582)$
  7. Approval seeking and sociability $(r = .581)$

III. *Essentially uncorrelated pairs: pairs in which less than 5 percent of the total variance is common variance.* $(r < .224)$
  1. Guilt and Cooperativeness $(r = .111)$
  2. Sociability and Creativity $(r = .047)$
  3. Initiative and Cooperativeness $(r = .072)$
  4. Approval Seeking and Adult Role $(r = .083)$
  5. Approval Seeking and Creativity $(r = .098)$
  6. Guilt and Sociability $(r = .022)$
  7. Approval Seeking and Initiative $(r = .051)$
  8. Mannerliness and Initiative $(r = .204)$
  9. Mannerliness and Guilt $(r = .204)$
  10. Mannerliness and Creativity $(r = .113)$
  11. Mannerliness and Adult Role $(r = .155)$
  12. Approval Seeking and Guilt $(r = .089)$
  13. Initiative and Sociability $(r = .123)$
  14. Creativity and Cooperativeness $(r = .120)$
  15. Sociability and Adult Role $(r = .201)$
  16. Adult Role and Cooperativeness $(r = .152)$

this something is not correlated with Mannerliness, Sociability, Co-operativeness, or Approval Seeking. One limitation is important here. There is no evidence—here—that Adult Role and Guilt are highly correlated. However, since they are not on the "low" list, they are not unrelated; thus we will go ahead with a tentative naming of this group. The traits all contain a connotation of the individual's "self-direction." This quality is required for creative activity and for carrying out the behaviors of an adult. Further, guilt implies—or could imply—self sanctions. Tentatively, then, we think, "'self-direction' is the name of whatever Adult Role, Initiative, and Creativity have in common."

It is apparent at this stage also that Mannerliness, Approval Seeking, Sociability, and Cooperativeness have something in common if only that each of them has nothing in common with members of our first threesome. Perhaps they represent "lack of self-direction". Pairs 3, 4, and 5 on the "high" list verify that three of the six possible pairs in this second grouping correlate highly.

This preliminary interpretation yields two groupings that are probably distinct. (The "probably" refers to the fact that not all possible pairs from within the groups are on the "high" list.)

The next step involves *revising our high and low criteria* to see what change, if any, this causes in our groupings. Let us now pick one-fourth, or 25 percent, of common variance as our division between high and low. We will now have a "moderately high" list for pairs such that $.250 \leq r^2 < .500$, and a "moderately low" list for pairs such that $.224 \leq r^2 < .250$. The "moderately high" pairs are shown in List III, Table 5.5.

The "moderately low" list would contain pairs correlated with r's that are significantly different from zero but that are not large in terms of common variance. There are no pairs on this latter list in Table 5.4.

Our interpretation from the first step can now be clarified. Item 5 on the "moderately high" list verifies that Guilt and Adult Role do correlate somewhat and belong in our original grouping. Items 1, 2, and 6 on the "moderately high" list suggest that Initiative might be interpreted as belonging to the first grouping, since it correlates moderately to all three other members. Furthermore, Initiative appears on the "low" list with the four variables in the second grouping.

The three pairs, of the six possible ones, in our second grouping showed up on the "moderately high" list. Therefore, we confirm and modify our first tentative groupings as follows.

Group I: Adult Role, Guilt, Creativity, and Initiative
Group II: Mannerliness, Sociability, Approval Seeking, and Cooperativeness

In each of Groups I and II, the variables have something in common—something not involved in the other group. For Group II this something might be "approval seeking" and for Group I this might be "self-regulation," a name that interprets Guilt as "self-punishment" and the other items as forms of "setting one's own goals."

It is important to be clear about what it means to say that variables in Group I are not related to variables in Group II. First, it has no reference to statistical significance. Some of our "low" correlations could be significantly different from zero: nevertheless, they account for too little of the total variance in the two variables to assist in interpreting the groupings.

Second, "not related" must be distinguished from "negatively correlated." For example, in Table 5.4 the correlation between Initiative and Cooperativeness is .072, or essentially, zero. This means that systematic relationship is absent; high ratings on Initiative are associated *equally as often* with "low," "moderate," and "high" ratings on Cooperativeness. This means that a child with much initiative may or may not engage in cooperative activities and that knowledge of a child's status on either of these traits will be of no value in predicting his status on the other. To further clarify this discussion, the next section will take up the interpretation of a table similar to Table 5.4 but with an important difference: two of the correlations are negative.

5.3.2 A Case with Some Negative r's

Table 5.6 has been prepared from Table 5.4 by changing the signs of two correlations and by deleting three of the variables to simplify

Table 5.6   *Modified intercorrelations among five of the traits from Table 5.4*

Trait	*Trait*				
	1. Mannerliness	2. Approval Seeking	3. Initiative	6. Creativity	7. Adult Role
1. Mannerliness	1.000	−.709	.204	.113	.155
2. Approval Seeking		1.000	.051	.098	.083
3. Initiative			1.000	.689	.582
6. Creativity				1.000	−.801
7. Adult Role					1.000

Table 5.7 *Interpretive lists of pairs of variables from Table 5.6*

I. *Higher correlated pairs: pairs in which more than 50 percent of the total variance is common variance.* ($|r| \geq .707$)

Mannerliness and Approval Seeking ($r = -.709$)
Creativity and Adult Role ($r = -.801$)

II. *Moderately high correlated pairs: pairs in which more than 25 percent, but less than 50 percent, of the total variance is common variance.* ($.500 \leq |r| < .707$)

Initiative and Creativity ($r = .689$)
Initiative and Adult Role ($r = .582$)

III. *Low correlated pairs: pairs in which less than 5 percent of the total variance is common variance.*

Mannerliness and Initiative ($r = .204$)
Mannerliness and Creativity ($r = .113$)
Mannerliness and Adult Role ($r = .155$)
Approval Seeking and Initiative ($r = .051$)
Approval Seeking and Creativity ($r = .098$)
Approval Seeking and Adult Role ($r = .083$)

interpretations. The interpretive lists for Table 5.6 appear in Table 5.7.

The changes have altered the function of the variables in defining each other and the groups. The "high" list suggests no clear groupings. However, the "low" list shows Initiative, Creativity, and Adult Role all essentially uncorrelated with Mannerliness and Approval Seeking, which in turn correlate negatively with each other. With the assistance of the "moderately high" list, we can suggest these groups:

Group I: Mannerliness and Approval Seeking
Group II: Initiative, Creativity, and Adult Role

The name of Group I must indicate the high negative correlation between the variables. That is, the name should suggest a trait such that children who are mannerly and don't seek approval have high ratings on it and children who are unmannerly and who seek approval have low ratings on it. Further, it must be unrelated to the items in Group II. Obviously this trait represents the degree of seeking approval by methods that ignore good manners. We could observe that infants and toddlers do this a lot, whereas older children tend to secure approval by conformity, including manners. Let us try "immature dependency" as a label for this group.

Group II contains three items that belonged to Group I in Table 5.4. However, before we rush in to label the something in common as "self-direction," let us remember that Creativity and Adult Role are negatively related. Perhaps high initiative can be expressed either in independent creative acts or in self-regulatory Adult Role behavior but not both. The term "independence" is safe: it suggests what all three variables have in common and still takes into account the very different kinds of independence assessed by the Creativity and Adult Role scores.

5.3.3 The Relevance of Specific Measures

In this section we will examine briefly a table of nonnegative correlations derived from Table 5.4 by altering some of the correlations involving Adult Role. Suppose that Adult Role were essentially uncorrelated with Initiative, Creativity, and Guilt, but that, instead, it were moderately to highly correlated with each of Approval Seeking, Mannerliness, and Cooperativeness. Such a state of affairs is reflected in Table 5.8, which is identical to Table 5.4 except for the seven correlations involving Adult Role.

An examination of the pairs, listed in Table 5.9, suggests that

Table 5.8 *Intercorrelations among eight hypothetical measures: modification of Table 5.4*

| Trait | Trait | | | | | | | |
	1. Mannerliness	2. Approval Seeking	3. Initiative	4. Guilt	5. Sociability	6. Creativity	7. Adult Role	8. Cooperativeness
1. Mannerliness		.709	.204	.081	.626	.113	.582	.774
2. Approval Seeking			.051	.089	.581	.098	.613	.652
3. Initiative				.671	.123	.689	.155	.072
4. Guilt					.022	.798	.063	.111
5. Sociability						.047	.801	.724
6. Creativity							.201	.120
7. Adult Role								.507
8. Cooperativeness								

Table 5.9 *Interpretive lists of pairs of variables from Table 5.8*

I. *Higher correlated pairs: pairs in which more than 50 percent of the total variance is common variance.* ($r \geq .704$)
    1. Adult Role and Sociability ($r = .801$)
    2. Guilt and Creativity ($r = .798$)
    3. Mannerliness and Cooperativeness ($r = .774$)
    4. Sociability and Cooperativeness ($r = .724$)
    5. Mannerliness and Approval Seeking ($r = .709$)

III. *Moderately high correlated pairs: pairs in which more than 25 percent of the total variance, but less than 50 percent of it, is common variance.* ($.500 \leq r \leq .704$)
    1. Initiative and Creativity ($r = .689$)
    2. Initiative and Guilt ($r = .671$)
    3. Approval Seeking and Cooperativeness ($r = .652$)
    4. Mannerliness and Sociability ($r = .626$)
    5. Mannerliness and Adult Role ($r = .582$)
    6. Approval Seeking and Adult Role ($r = .613$)
    7. Approval Seeking and Sociability ($r = .581$)
    8. Adult Role and Cooperativeness ($r = .507$)

IV. *Essentially uncorrelated pairs: pairs in which less than 5 percent of the total variance is common variance.*
    1. Guilt and Adult Role ($r = .063$)
    2. Initiative and Adult Role ($r = .155$)
    3. Creativity and Adult Role ($r = .201$)
    4. Guilt and Cooperativeness ($r = .111$)
    5. Sociability and Creativity ($r = .047$)
    6. Initiative and Cooperativeness ($r = .072$)
    7. Approval Seeking and Creativity ($r = .098$)
    8. Guilt and Sociability ($r = .022$)
    9. Mannerliness and Initiative ($r = .051$)
    10. Mannerliness and Guilt ($r = .081$)
    11. Initiative and Sociability ($r = .123$)
    12. Approval Seeking and Guilt ($r = .089$)
    13. Initiative and Sociability ($r = .123$)
    14. Creativity and Cooperativeness ($r = .120$)
    15. Mannerliness and Initiative ($r = .204$)

Initiative, Guilt, and Creativity constitute one group. This interpretation is based on the fact that each of the named variables correlates with each other one and that all three of them have "low" correlation with Mannerliness, Adult Role, and Cooperativeness. The latter three variables are themselves intercorrelated. The groupings for Table 5.8 appear, then, to be

Group I: Adult Role, Mannerliness, Approval Seeking, and Co-
    operativeness
Group II: Initiative, Creativity, and Guilt
Group III: Sociability

Group I emerges clearly only when the "moderate" list in Table
5.9 is considered. At the first stage ("high" and "low" lists only),
the absence of correlation ("low" list) is more suggestive of the
final groupings than is the presence of correlation ("high" list).

In any case, observe that the three groups are the same as those
for Table 5.5 *except* for the location of Adult Role in Group I here
and in Group II formerly. Group I was formerly (without Adult Role)
called "approval seeking." This term may still be appropriate if we
interpret Adult Role as a behavior exhibited by children in order to
obtain adult approval.

On the other hand, we may wish to retain our original interpretation of Adult Role. It was that taking an adult role facilitates
self-direction; one needs less adult guidance (for example, approval)
if one can guide oneself, using behavior often seen in adults. If this
is so, we would have to change the label used for Group I to something
like "maturity," indicating that using good manners and cooperating
were instances of advanced or mature behavior. This usage would
require that "approval seeking" refer to getting peer approval or
something else that is more advanced than a child's reliance on adult
approval.

Under either interpretation of Adult Role, Group II might still
well be called "self-regulation." The important lesson of this section
is that interpreting many correlations requires judgment. In the
change from Table 5.4 to Table 5.8, there is a clear *choice* as to whether
to alter one's view of the meaning of the variable Adult Role or
whether to alter one's view of the meaning of the groupings. The
groupings are interpreted in part through reference to member
variables; member variables are, in turn, interpreted in part through
reference to groupings. The use of several mathematical models to
aid the interpretive process will be discussed in the next sections.
The careful reader will observe while studying these methods that
individual reader judgment retains its crucial role in the interpretation of the results of multivariate research.

It should be noted that the more formal technique of factor analysis (discussed in Section 5.5) is often used to discover the underlying
structure of correlation matrices such as those we have been discussing in this section. The technique of this section is uniquely appropriate for preliminary, "intuitive" reader interpretation of correla-

tion tables that appear in journal articles without the corresponding factor-analytic results. It should also be noted that the process discussed in this section of finding a name for the groups of variables has much similarity to the factor-naming process discussed in Section 5.5.

5.4 Multiple Linear Correlation and Regression

When the investigator's purpose is to predict one variable from two or more other variables, the techniques of multiple-regression analysis are appropriate. The variable to be predicted is termed the *dependent variable* and the several predictor variables are termed *independent variables*. These terms do not imply that the predictor variables are manipulated by the investigator; usually they are not. Rather, the terms are adopted to conform with mathematical usage. In research reporting, "criterion" and "predictor" are often substituted for "dependent" and "independent," respectively. Given a dependent variable, the linear-regression problem is to estimate constants B_1, B_2, ... B_k, and A such that the expression

$$Y = B_1X_1 + B_2X_2 + \cdots + B_kX_k + A \qquad (5\text{-}6)$$

provides a good estimate of an individual's Y score based on his X scores.

In most reports and discussions, Y and the several X variables are converted to standard scores; z_Y, z_1, z_2, $z_3 \ldots z_k$; each z has a mean of 0 and standard deviation of 1. Then, the problem is to estimate constants, β_i, such that

$$z'_Y = \beta_1 z_1 + \beta_2 z_2 + \cdots \beta_k z_k \qquad (5\text{-}7)$$

Here z'_Y stands for the predicted value of the standardized Y score, z_Y. The expression to the right of the equals sign is referred to as a linear combination of the independent variables. Hence the term "multiple *linear* regression." The constant A is eliminated in the process of converting X's to z's.

The least-squares method is used to estimate the beta weights. The calculations are rarely shown in a research report, but the general idea follows:

Consider the difference, $z_Y - z'_Y$, between an individual's z_Y score and the one predicted for him by Equation 5-7. Call this difference the "prediction error" for that individual. The least-squares method finds β's in such a way that the sum of the squared prediction errors is kept as small as possible; that is, in such a way that the expression $\Sigma(z_Y - z'_Y)^2$ is minimized. The details of the least-squares method include first the use of differential calculus and, subsequently, the algebraic solution of k equations in k unknowns (the beta weights). The predictive adequacy of a set of beta weights is indicated by the size of the correlation coefficient $r_{z_Y z'_Y}$ between the predicted z'_Y scores and the actual (measured) z_Y scores. This special Pearson correlation coefficient is called the *multiple correlation coefficient*, and is often symbolized by R.

The squared multiple correlation, R^2, represents the proportion of criterion (z_Y) variance accounted for by the predictors, that is, the proportion of total variance that is common variance.

For the reader, the most useful thing to know about prediction equations (and their associated multiple-correlation coefficients) is that both R and the beta weights depend on the pairwise correlations among all the predictor variables. The correlation between a criterion variable and a potential predictor is not a direct indication of how valuable that predictor will be as one among several. To develop this insight, let us explore in detail an example from Table 5.4 (page 134). After that, we will summarize the principles for the general case.

Suppose the investigator wishes to predict Creativity scores for 100 individuals, and he has at his disposal the 700 scores for these individuals on the remaining seven variables in Table 5.4. Further, let us assume that the correlations in Table 5.4 were computed from the scores of 1000 *other* individuals[4] by using all seven available predictors in one equation. The beta weights and resultant prediction equation follow.

$$z'_6 \quad = \quad -.081\,z_1 \quad + \quad .098\,z_2 \quad + \quad .184\,z_3$$

Creativity   Mannerliness   Approval Seeking   Initiative

$$+ \frac{.360\,z_4}{\text{Guilt}} - \frac{.173\,z_5}{\text{Sociability}} + \frac{.495\,z_7}{\text{Adult Role}} + \frac{.116\,z_8}{\text{Cooperativeness}} \qquad (5\text{-}8)$$

---

[4] The beta weights will constitute the least-squares solution *only* for the 1000 *other* individuals. Slightly different betas would result for the 100 individuals the investigator is studying. The situation presented is, however, the one most often seen in practice. If the investigators had "creativity" scores for his 100 subjects, he would not need to predict them! The reader should know that R is usually somewhat smaller for a second sample when it is computed using z'_Y predicted from the regression equation for the first sample. This sample-to-sample shrinkage is inversely related to the number of variables and directly to the number of cases in the first sample. In this example the first sample is sufficiently large to make R from the 1000 cases in Table 5.4 a good estimate of R for the 100 cases at hand.

Each β_i is proportional to the correlation between the criterion and that part of the predictor, z_i, that is independent of the remaining predictors. Such a correlation is called a part correlation. Thus a given beta, β_i, is defined like this:

$$\beta_i = \frac{\text{part } r}{1 - R_i^2} \qquad \text{(between } Z_6 \text{ and the independent part of } z_i\text{)}$$

where R_i^2 is the multiple-correlation coefficient associated with predicting z_i from the other six predictors. Thus the individual beta weights tell us directly how much each predictor contributes to the multiple-correlation *independent* of its fellow predictors. R^2 represents the total criterion variance "explained," and β_i^2 is that portion of explained variance independently contributed by z_i. It is important to note that the sum of the squared betas, $\Sigma\beta_i^2$, *is not* equal to R^2. In our example, $R^2 = .776$ and $\Sigma\beta_i^2 = .708$; this is because some variance in the criterion is explained by a joint contribution from two or more predictors acting together.

An examination of Equation 5-8 and several others will serve to illustrate the process of interpreting beta weights. In Equation 5-8, we note that the largest beta weights are attached to the three variables that grouped with the criterion in our initial examination of Table 5.4 (Section 5.3.1). These three variables have the highest correlations to the criterion, so their beta weights are intuitively reasonable. It appears that each of z_3, z_4, and z_7 continues to correlate (but with not nearly so high a correlation) when the influence of other predictors is removed.

Consider, specifically, $\beta_3 = .184$. The simple correlation between z_3 (initiative) and the criterion z_6 (creativity) is found in Table 5.4 as .689. How can we relate this .689 to $\beta_3 = .184$? Substituting β_3 in Equation 5-9 we see that

$$.184 = \frac{\text{part } r}{1 - R_3^2} \qquad (5\text{-}9)$$

which implies that β_3 is small relative to $r_{6\,3}$, either because the part r between z_6 and the part of z_3 independent of the others is small *or* because the multiple correlation R_3^2 is small, thus making the denominator of Equation 5.9 large.

We are usually not given values for the part correlations or the several multiple R's among the predictors, but we can look at what happens to the beta weights when only certain predictors are used. Let us compare the following combinations of predictors.

1. The best single predictor, z_7:

$$z'_6 = .801 \, z_7 \tag{5-10}$$

$$R^2 = .642$$

2. The two best predictors from the same group:

$$z'_6 = .492 \, z_y + .500 \, z_7 \tag{5-11}$$

$$R^2 = .890$$

3. All the predictors from the same group:

$$z'_6 = .136 \, z_3 + .423 \, z_4 + .462 \, z_7 \tag{5-12}$$

$$R^2 = .895$$

It is clear that something is gained by adding one additional predictor to the best single predictor; however, very little is gained by adding the third predictor. Indeed, notice that the β_3 for Equation 5-8 is not much different from the β_3 for the equation that uses only two predictors from its group, Equation 5-11.

The general principle illustrated in the example is that when predictors are highly correlated among themselves, there is little predictive value (increase in R^2) gained by adding more than two predictors to the regression equation. Actually, Table 5.4 is not ideal for purposes of prediction. There is too much redundancy, which means that one predictor provides much the same information as the other predictors from its group. This leads to wasteful duplication.

Prediction by means of multiple linear regression is most successful when (a) the intercorrelations among predictors is low[5] and (b) the correlation between each predictor and the criterion is moderate. Under this circumstance, the independent contribution of each predictor is optimized. If the correlations among predictors is high, as in Equation 5-12, then each to some extent duplicates the effect of the other. This fact is then reflected in smaller beta weights for each variable and a higher portion of R derived from the joint contribution.

[5] Occasionally what is known as a "suppressor variable" is encountered as a predictor. This type of variable *does* correlate well with another predictor, but not with the criterion. Yet its use *increases* R^2. This happens because the correlation of the other predictor with the criterion (the part correlation) is increased when the suppressor is removed. Such situations are rare in practice.

To get a better idea of the independent contribution of each predictor, some investigators add them into the equation one by one, computing betas and R^2 at each step. This procedure is illustrated by the progression from Equations 5-10 to 5-12. Formal computerized techniques are available for admitting variables only if they increase the predicted variance, R^2, by a specified amount, say, .05. These techniques are known as *stepwise-regression* techniques and can be helpful in aiding the investigator to identify the specific contribution of each variable. This specific contribution can be better assessed by reference to R^2 changes than by reference to beta weights. If this information is not available, then you will need to study the entire equation and check for (1) the relation of r_{Yi} to β_i, (2) the amount of joint contribution to R^2 that is indicated by the discrepancy between

$$\sum_{i=1}^{k} \beta_i^2$$

and R^2, and (3) the various intercorrelations among the predictors.

Finally, it should be mentioned that the null hypothesis (that $R = 0$ in the population) can be tested by means of one-way analysis of variance; this procedure is often seen in the literature and is legitimate. However, for R, as for r, we shall retain the view that the magnitude of R^2, the proportion of predicted variance, is more useful for interpretation than is the fact that R is nonzero. This is especially so if some recommendations for action (graduate school admission, new teaching technique, job assignment) is going to be based on the regression equation. Small but significant R's may have value in suggesting lines of future investigation and the like, but they are of little immediate practical value.

5.5 Factor Analysis

Factor analysis is a technique that is applied to tables of pairwise correlations, such as Table 5.4 of Section 5.3. The purpose is to reduce the number of variables to just one, two, or several basic ones; these basic variables are called *factors*. Once defined, factors function in the same way as the names of the groupings in Section 5.3. A given factor, then, represents whatever it is that a group of one or more variables has *in common*. The term *common factor* is often used to distinguish a factor common to two or more simple variables from a factor present in only one variable in the set. The latter is called a *specific factor*.

A given matrix of correlations among a number, k, of variables necessarily has k specific factors and usually has fewer than k common factors. There may be only one common factor, a *general factor;* in which case, all variables have some one attribute in common. There may be two or three or more factors. If so, the variables would divide themselves into groups; each group would have something in common, namely, the common factor that defines the group. This might seem appropriate when the variables are aptitude-test scores and there are several main aptitudes present in a large number of tests. Another possibility would be to have (*a*) one general factor common to all variables and also (*b*) some common factors that define groups of variables. This latter possibility often seems appropriate in factor analysis with mental tests, since one can postulate some general factor called "intelligence" and several common factors for specific mental traits such as "memory," "vocabulary," "abstract reasoning," and so forth. In some types of factor analysis, variables may enter into one or more groupings; in other models, this possibility is ruled out.

In Section 5.5.1, the process of interpreting factor analysis results will be illustrated. There exist a number of different models for factor analyses, all of which provide different results (for the same data). Once you have studied the interpreting process, you will become acquainted with the general mathematical rationale (Section 5.5.3), the effect of different assumptions (Section 5.5.4), and the relation of these matters to your task of critical interpretation (Section 5.5.5).

5.5.1 *Interpreting the Results of a Factor Analysis*

Once a factor analysis has been performed, the investigator presents his results as shown in Table 5.10a; such a table is often called a "factor matrix." (This term will be explained in a later section.) Let us here define and study the parts of Table 5.10a. Each entry in columns one and two represents the correlation between the row variable and the column factor. Thus the variable Sociability has a correlation of $r_{5I} = .56$ with Factor I, and $r_{5II} = .54$ with Factor II. These special correlations between variables and hypothetical factors are called *factor loadings*. Table 5.10 has 16 factor loadings, one loading for each variable on each factor.

The entries in the right-hand margin are called *communalities*. Each communality represents the proportion of variance in the corresponding (row) variable and is accounted for by the two common factors. Thus 75 percent of the variance in Mannerliness is accounted for by Factors I and II. The remaining 25 percent of the total variance in Mannerliness scores is thought of as being made up of two parts: a factor specific to the Mannerliness attribute and a portion due to errors of measurement involved in the assessment of Mannerliness. There is no representation in Table 5.10a of these portions of the

Table 5.10a *Principal factors in Table 5.4*

Variable	Factor I	Factor II	Communality
1. Mannerliness	.65	.57	.75
2. Approval Seeking	.54	.54	.58
3. Initiative	.61	−.45	.57
4. Guilt	.63	−.54	.69
5. Sociability	.56	.54	.61
6. Creativity	.72	−.59	.87
7. Adult Role	.67	−.45	.65
8. Cooperativeness	.64	.60	.77
Common variance	3.17	2.31	5.49
Proportion of total variance	.40	.29	.69
Proportion of common variance	.58	.42	1.00

Table 5.10b *Principal factor solution for the correlation matrix of Table 5.4 with varimax rotation*

Variable	Factor I	Factor II	Communality
1. Mannerliness	.10	.86	.75
2. Approval Seeking	.04	.77	.59
3. Initiative	.76	.08	.58
4. Guilt	.83	.03	.69
5. Sociability	.05	.77	.60
6. Creativity	.93	.05	.87
7. Adult Role	.80	.11	.65
8. Cooperativeness	.07	.87	.76
Variance accounted for	2.79	2.70	5.49
Proportion of total variance	.35	.34	.69
Proportion of common variance	.51	.49	1.00

Note: See page 164 for a discussion of the varimax rotation.

Table 5.10c   *Principal component solution for the correlations of Table 5.4*

Variable	Principal Components								
	1st	*2nd*	*3rd*	*4th*	*5th*	*6th*	*7th*	*8th*	*h²*
1. Mannerliness	+.68	+.58	−.17	+.14	−.17	−.28	+.18	+.09	.954
2. Approval Seeking	+.60	+.59	−.33	−.25	−.18	+.28	−.06	−.06	1.056
3. Initiative	+.65	−.52	−.13	+.50	−.11	+.08	−.11	−.04	.993
4. Guilt	+.65	−.59	−.23	−.09	+.35	+.07	+.19	−.04	.998
5. Sociability	+.61	+.57	+.41	+.16	+.12	+.28	+.06	+.10	.998
6. Creativity	+.71	−.61	.01	−.19	−.02	−.05	−.16	+.23	.993
7. Adult Role	+.69	−.49	+.37	−.21	−.24	−.07	+.08	−.14	.996
8. Cooperativeness	.67	−.61	+.06	−.04	+.26	−.24	−.18	−.11	1.004
Proportion of Total variance	.44	.33	.05	.05	.05	.05	.02	.01	1.000

Table 5.10d   *Varimax rotation of the components in Table 5.10c*

Variable	Principal components								
	1st	*2nd*	*3rd*	*4th*	*5th*	*6th*	*7th*	*8th*	*h²*
1. Mannerliness	.00	.30	.38	.11	.05	.82	.27	.00	.99
2. Approval Seeking	.05	.27	.90	.00	.02	.28	.18	.01	1.05
3. Initiative	.33	.05	.00	.88	.32	.09	−.01	.04	.99
4. Guilt	.88	−.02	.04	.30	.36	.00	.04	.02	1.00
5. Sociability	−.03	.91	.26	.05	.08	.23	.22	−.01	1.00
6. Creativity	.50	−.04	.06	.33	.67	.01	.06	.43	.99
7. Adult Role	.24	.11	.01	.22	.94	.05	.03	−.04	1.00
8. Cooperativeness	.07	.43	.32	−.02	.05	.39	.74	.02	1.00

variance; this is because the focus in factor analysis is on the common variance.

In the example of Table 5.10a (but not always), the factors are considered to be uncorrelated. As a direct consequence of this fact, the communality of each variable can be obtained as the sum of the squares of its loadings.[6] Thus the communality for Mannerliness, symbolized h_1^2 is found as $h_1^2 = (.65^2) + (.57^2) = .75$ or, in general, for the jth variable in a set with p factors:

$$h_j^2 = r_{jI}^2 + r_{jII}^2 + \cdots r_{jp}^2 \tag{5-13}$$

Notice that the H^2 used earlier is for the common variance in the

[6]See Section 5.5.3.

entire group of variables, whereas h_i^2, which is used here, represents that portion of one variable's variance that is common to all of the others. Equation 5-13 is exactly analogous to the formula for R when p predictors are uncorrelated, which is

$$R^2 = \beta_1^2 + \beta_2^2 + \cdots \beta_p^2 \qquad (5\text{-}14)$$

In both cases, the left-hand symbol (h_j^2 or R_j^2) represents the proportion of variance in a jth variable that may be accounted for by a linear combination of the variables associated with the terms on the right.

From this discussion, we see that the variables in a factor analysis can be considered as criterion variables to be predicted. Instead of predicting one variable from the best linear combination of the remaining variables, we predict each of them from the factors. Thus we can conceive of (and find) scores on the factors for the individuals.

The *important distinction* here is this: in the multiple-regression problem, predictor scores for individuals are obtained directly through a known measuring instrument such as a test or rating scale. In factor analysis, the predictors are hypothetical unknown factors, and factor scores are derived (in a mathematically complex way) from the scores on the criterion variables. Factors are determined in such a way as to necessarily make each of them a good predictor of all or some subset of the variables.

Each factor in Table 5.10a (and any factor table) is defined by examining the pattern of its loadings. The names and meanings of the variables are employed to give meaning, and a name, to the factors. Since the loadings may be interpreted as correlation coefficients, some arbitrary criterion of "great enough" must be adopted for deciding when a variable is to be included in the list of those used to interpret a factor. It is customary in factor-analysis literature for a loading of .33 to be the minimum absolute value to be interpreted. The portion of a variable's variance accounted for by this minimum loading is approximately 10 percent. This criterion is arbitrary and is used here to conform to the practice likely to be encountered by the reader of multivariate research reports.

The first factor in Table 5.10a has loadings in excess of .33 on all of the variables; such a factor is usually called "the general factor" and is taken to represent whatever it is that all of the variables have in common. We might consider all eight traits to be the product of the socializing process. Hence a term like "social maturity" could be used to indicate some quality that is present in varying degrees in all individuals in the group and which is part of all the traits measured. Other terms like "developmental status," "internalization," or "socialization" would do as well.

The factor name is chosen to convey as well as possible what it is that all the variables that correlate with it (that "load on it") have in common. Factor II also has all loadings in excess of .33; however, half of them are negative. Such a factor is called a bipolar factor and is taken to represent a single dimension with two poles. Each of these poles is defined by a cluster of variables—one pole by those with positive loadings and one pole by those with negative loadings. Notice that, in this case, the positive pole is to be defined by the variables Mannerliness, Approval Seeking, Sociability, and cooperativeness, which constituted Group II in Section 5.3.

The negative pole of Factor II, similarly, is to be defined by the remaining four variables, which constitute Group I from Section 5.3. We can, perhaps, use the names of Group I ("self direction") and Group II ("approval seeking") to help us interpret and name Factor II. An individual who has a high degree of "Factor II-ness" is one who is self-directed and who does not seek approval in the manner connoted by the variables of Group I. An individual low in "Factor II-ness" is one who seeks approval by being cooperative, mannerly, and sociable—and who is not self-directed. Since manners, cooperation, and sociability all require the presence of other people, and self-direction does not, it is tempting to call this factor the "other-directed—self-directed" dimension. There are probably a dozen other terms that would do as well; perhaps the reader can think of several.

The rows at the bottom of Table 5.10a give us further information about the usefulness of the two factors in explaining the relations among the eight variables. The total variance in the analysis can be symbolized V. If the variables are standardized, each one has an individual variance, $\sigma_j^2 = 1.00$, and thus

$$V = \sum_{j=1}^{8} \sigma_j^2 = 8.00$$

The row labeled "common variance" gives the numerical value of that portion of the variance attributed to the column above it. These variances are found as

$$V_p = \sum_{j=1}^{8} r_{jp}^2$$

Thus the total value, 8.00, is partitioned into $V_1 = 3.17$ for Factor I; $V_2 = 2.31$ for Factor II; and $V_c = 5.48$, the total for Factor I and Factor II. The corresponding proportions of the total variance, 8.00, are shown in the next row; there we can see that 69 percent of the total variance, $V = 8.00$, is related to these two factors.

Thirty-one percent of the total variance is accounted for by specific factors and error of measurement variance. Approximately 69 percent of the total variance in Table 5.4 (The original correlations) is *common variance* whereas 31 percent of it is made up of portions unique to individual variables and the techniques employed to assess them. The last row shows that, *of the common variance*, approximately 58 percent is accounted for by Factor I and the other 42 percent by Factor II. The 58 percent is found as $100 \times 3.17/5.49$; that is, it represents the percent of common variance, not the percent of total variance. The 42 percent is found similarly; thus it can be concluded that the two factors together "explain" the common variance.

In summary, the results of a factor analysis consist in a new table of correlations with k rows and p columns; the entries are correlations between variables and the factors they define. Factors are interpreted by observing the variables that correlate with them and the signs of those correlations. The principal-factor method, here illustrated, produces first (a) a general factor common to all variables and then (b) one or more bipolar factors depicting dimensions whose high ends are characterized by one subgroup of variables and whose low ends are characterized by another.

5.5.2 *The Algebraic Rationale of Factor Analysis. (Optional Section)*

The interpretation of a completed factor analysis derives from the individual entries in the cells of factor matrix—called "loadings"—which are, in fact, correlations between variables and factors. The calculation of correlations requires two sets of scores; therefore, factors must be traits that can be given scores. The basic factor-analysis model states that each variable is a linear combination of the factors. This model can be written as follows for variable j and all factors in standard-score form:

$$z_j = r_{j\mathrm{I}}F_\mathrm{I} + r_{j\mathrm{II}}F_\mathrm{II} + \cdots r_{jp}F_p + U_j \tag{5-15}$$

Here $F_\mathrm{I}, F_\mathrm{II} \ldots F_p$ represent the standard score value of the common factors and U_j represents that portion of z_j that is not common to the other variables.

The values of the loadings are found as solutions to simultaneous equations. Each equation expresses a correlation r_{jk} (known) between two of the original variables in terms of the factor loadings r_{jp}

(unknown). Since there are $\frac{1}{2}k(k-1)$ possible correlations among k variables, there are this same number of equations. Let us first derive the expression for one such equation, say the one for X_1 and X_3. First we write the correlation formula in standard-score form (see Chapter 3):

$$r_{13} = \frac{\sum\limits_{i=1}^{N} (z_{i1} z_{i3})}{N} \tag{5-16}$$

Then we express each of z_1 and z_3 in terms of the factors using the expression in Equation 5-15:

$$r_{13} = \frac{\sum (r_{1\text{I}}F_{\text{I}} + r_{1\text{II}}F_{\text{II}} + \cdots r_{1p}F_p + U_1)(r_{3\text{I}}F_{\text{I}} + r_{3\text{II}}F_{\text{II}} + \cdots V_{3p}F_p + U_3)}{N}$$

The summation is taken over individuals, i, although this subscript has been omitted. By performing the indicated multiplication and extending the summation sign, the expression may be rewritten as

$$r_{13} = r_{1\text{I}}r_{3\text{I}} \frac{\sum F_1^2}{N} + r_{1\text{II}}r_{3\text{II}} \frac{\sum F_{\text{II}}^2}{N}$$

$$+ \cdots r_{1p}r_{3p} \frac{\sum F_p^2}{N} + r_{3\text{I}}r_{3\text{II}} \frac{\sum F_{\text{I}}F_{\text{II}}}{N}$$

$$+ \cdots r_{1p}r_{3(p-1)} \frac{\sum F_p F_{p-1}}{N} + r_{3p} \sum F_p U_1 + r_{1p} \sum F_p U_3 + \frac{\sum U_1 U_3}{N}$$

All $\Sigma F_p^2/N$ are 1.000, since the F's are standard scores. The sums of products terms like $F_{ip}F_{3p}/N$ represent correlations. A necessary assumption for solutions is that all unique parts U_j are uncorrelated to each other or to common factors. Thus Equation 5-16 becomes:

$$r_{13} = \underbrace{r_{1\text{I}}r_{3\text{I}}} + \underbrace{r_{1\text{II}}r_{3\text{II}}} + \cdots + \underbrace{r_{1(p-1)}r_{3p}} + \underbrace{r_{1p}r_{3(p-1)}} + \underbrace{r_{1p}r_{3p}}$$

Products of factor
loadings (unknown)

$$+ \underbrace{r_{1p}(r_{1\text{I}}r_{3p} + r_{1p}r_{3\text{I}})} + \cdots + \underbrace{r_{(p-1)p}} \tag{5-17}$$

Correlations between
pairs of factors (in
many solutions assumed
to be 0)

There are $\frac{1}{2}k(k-1)$ equations such as Equation 5-17. When factors are uncorrelated these equations can be written as:

$$r_{jk} = r_{jI}r_{kI} + r_{jII}r_{KII} + \cdots + r_{jp}r_{kp} \qquad (5\text{-}18)$$

The expression in Equation 5-18 provides for $\frac{1}{2}k(k-1)$ equations in kp unknowns. If the factors are correlated, there are $\frac{1}{2}p(p-1)$ more unknowns, one for each pairwise correlation among factors. In any case, the equations are solved by methods of matrix algebra, usually with the aid of a high-speed electronic computer. The reader of a report will be presented with the finished solution in the form of a factor matrix such as Table 5.10a.

5.5.3 Problems in Factor Analysis and the Models that Represent Different Solutions to Those Problems

A factor matrix such as Table 5.10a may be interpreted with the help of suggestions given in the preceding section. But several problems are encountered in working out the previously mentioned algebraic solution for the loadings. The reader should be aware that these problems exist; that each presents the investigator with several decisions to make; and especially, that an investigator's results—that is, the values and patterning of the loadings—will be dependent upon his decisions as well as on the structure in his data.

The problems that arise in finding a factor-analysis solution (like Table 5.10a) for a given correlation matrix (like Table 5.4) all relate to a single algebraic fact about simultaneous equations: In order for there to be one and only one solution (that is, a unique solution), there must be the same number of variables as there are equations. Otherwise, there may be *several* sets of values (r_{jp}) that would serve equally well from the viewpoint of mathematics.

In factor analysis, there are $\frac{1}{2}p(p-1)$ unknowns. The constant k, the number of variables, is known. The constant p is, however, unknown. This means that the number of unknowns—the number of factors—is itself a problem for solution (that is, an unknown).

5.5.3.1 Communality and the number of factors. Logical considerations suggest that p should be small relative to k in order that the factor analysis shall serve its purpose of reducing complexity; however, if p is in fact small relative to k, (a) there will be fewer unknowns (loadings) than there are equations, (b) a unique solution for the loadings cannot be found. To provide the basis for an initial algebraic solution, the following assumptions are made.

1. The factors are uncorrelated; thus, all interfactor r's $= 0$ in Equation 5-17.
2. There are $p = k$ factors.

These assumptions permit the solution to begin. The factors are extracted ("solved for") one by one; the factoring is stopped when there is attained a number of factors (usually less than k) adequate to account for the common variance among the variables.

This poses a new problem: how is the amount of common variance to be known in advance of solution? In Section 5.5.2, an example of the results of a factor analysis was given (see Table 5.10a), and the common variance h_j^2 for each variable x_j was determined by summing the squares of the factor loadings. It was shown that, given p, the number of factors, the common variance,

$$H^2 = \sum_{j=1}^{k} h_j^2$$

can be found. Now, in the previous paragraph it was suggested that, given a known amount of common variance, the number, p, of factors can be found. This is because there is a circular relationship between total communality and the number of factors; this circular relationship is clarified in the expression from Table 5.10a:

$$H^2 = \sum_{l=1}^{p} \sum_{j}^{k} r_{jl}^2 \tag{5-19}$$

The summation is taken over loadings for each variable on each factor. The value of p can be chosen so that the indicated sum will equal a previously estimated H^2 value; or p can be specified, the solution can be carried out, and *then* H^2 can be found.

5.5.3.2 Solutions: Principal Factors and Principal Components. Three approaches have been used to solve the number-of-factors and communality-circularity problem. These approaches amount, in practice, to different answers to the question, "What entries go into the diagonals of a correlation table before it is factor analyzed?"

1. *Assume, on theoretical grounds (concerning the nature of the variables), that $p = 1$ or $p = 2$.* For these assumed values, there exist estimates, h_j^2, for each variable. These h_j^2 estimates are placed in the diagonals (representing $r_{11}, r_{22} \ldots r_{jj}$) of the correlation matrix when it is subjected to factor-analysis solution. The solution for the first

one or two factors will necessarily recover the original communality H^2 as estimated.

2. *Estimate h_j^2 on some basis, other than Equation 5-18, that is not dependent on the number of factors.* Some commonly suggested estimators for h_j^2 are

(a) The squared multiple-correlation coefficient (SMC) between z_j as criterion and all the other variables as predictors.

(b) The greatest correlation between z_j and any *one* of the remaining variables.

(c) The reliability of variable j as estimated from conventional psychometric formulae.

3. Do not try to estimate either h_j^2 or p, the number of factors. Instead of accounting for the common variance, H^2, account for all the variance V, and extract the $p = k$ factors to do the job. (A unique solution to the simultaneous equations exists in this case if the diagonal cells contain 1.00's.)

The first solution was used historically in the first factor analyses of the scores obtained from mental tests. A factor of general intelligence, g, was considered to exist and a one-factor factor analysis was performed to identify it. This approach is of little value today, since the computer has made the other two feasible and they both allow for the possibility of a single factor.

The next solution is often used today. That is, the communalities are estimated by squared multiple correlations $R_j^2 = h_j^2$, and then the matrix solution proceeds until enough factors emerge to account for the estimated total communality, H^2. The most common matrix method in this instance is called the *principal-factor solution.* An important property of this solution is that the first factor found is the one that accounts for the greatest proportion of common variance; the second factor found accounts for the next greatest proportion of common variance, and so on. When the common variance, as estimated, is accounted for as nearly as possible, the solution is stopped.

Tables 5.12b, 5.13b, and 5.14b illustrate principal-factor solutions for the three different three-variable situations that were first illustrated in Tables 5.1, 5.2, and 5.3. Table 5.11 gives the general case. In Table 5.12a the variables are virtually unrelated. The total common variance H^2 estimated by totaling the three multiple R's is .049, which, when divided by the total variance 3.00, gives a proportion .016. Only 1.6 percent of the variance is common variance. The first factor accounts for more than this proportion of variance and shows no loading greater than .300. Thus, there is no common factor—which makes sense, since there is no common variance.

Table 5.11 *Principal factors: generalized table*

Variable	Factor I	Factor II	$\cdots$	Factor p	Communality
x_1	r_{1I}	r_{1II}		r_{1p}	$r_{1I}^2 + r_{1II}^2 + \cdots + r_{1p}^2$
x_2	r_{2I}	r_{2II}		r_{2p}	$r_{2I}^2 + r_{2II}^2 + \cdots + r_{2p}^2$
x_3	r_{3I}	r_{3II}		r_{3p}	$r_{3I}^2 + r_{3II}^2 + \cdots + r_{3p}^2$
x_4	r_{4I}	r_{4II}		r_{4p}	
$\cdots$	$\cdots$	$\cdots$		$\cdots$	$\cdots$
x_k	r_{kI}	r_{kII}		r_{kp}	$r_{kI}^2 + r_{kII}^2 + \cdots + r_{kp}^2$
Common variance	$\sum_{j=1}^{k} r_{jI}^2$	$+ \sum_{j=1}^{k} r_{jII}^2 \cdots$	$+$	$\sum_{j=1}^{k} r_{jp}^2$	$= \sum_{l=1}^{p}\sum_{j=1}^{k} r_{jl}^2$
Proportion of total variance	$\sum_{j=1}^{k} r_{jI}^2/k$	$+ \sum_{j=1}^{k} r_{jII}^2/k \cdots$	$+$	$\sum_{j=1}^{k} r_{jp}^2/k$	$= \sum_{l=1}^{p}\sum_{j=1}^{k} r_{jl}^2/k$
Proportion of common variance	$\sum_{j=1}^{k} r_{jI}^2 \Big/ \sum_{l=1}^{p}\sum_{j=1}^{k} r_{jl}^2$	$+ \sum_{j=1}^{k} r_{jII}^2 \Big/ \sum_{l=1}^{p}\sum_{j=1}^{k} r_{jl}^2 \cdots$	$+$	$\sum_{j=1}^{k} r_{jp}^2 \Big/ \sum_{l=1}^{p}\sum_{j=1}^{k} r_{jl}^2$	$= 1.000$

Table 5.12 *Three unrelated variables: two analyses*

a. correlation matrix

	IQ	Height	Music Aptitude
IQ	.008*	.032	.092
Height	.032	.017*	.128
Music aptitude	.092	.128	.024*

b. Principal factor solution

Variable	Factor I, (r_{jI})	Communality, (h_j^2)
x_1 (IQ)	.195	.038
x_2 (Height)	.254	.064
x_3 (Music aptitude)	.300	.090
Variance accounted for	.192	.192
Proportion of total variance	.064	.064
Proportion of common variance	1	1

c. Principal components

Variable	Component I	Component 2	Component 3	Communality
IQ	.501	−.805	.316	1.000
Height	.628	.562	.538	1.000
Music aptitude	.727	.067	−.683	1.000
Variance accounted for	1.175	.968	.857	3.000
Proportion of total variance ($V = 3$)	.391	.324	.285	1.000

d. Principal-component solution, varimax rotation

Variable	Factor I	Factor II	Factor III	Communality
x_1 (IQ)	.015	.999	.046	1.000
x_2 (Height)	.998	.015	.064	1.000
x_3 (Music aptitude)	.064	.046	.997	1.000
Variance accounted for	1.000	1.000	1.000	$3.000 = V$
Proportion of total variance	.333	.333	.333	1.000

*Squared multiple R, which is replaced by 1.000 for the principal-component solution in part c.

Table 5.13 *Three variables, two related: two analyses*

a. Correlation matrix (From Table 5.2)

	IQ	Height	College Aptitude
IQ	.814*	.032	.902
Height	.032	.001*	.036
College aptitude	.902	.036	.814*

b. Principal-factor solution

Variable	Factor I	Communality
x_1 (IQ)	.926	.857
x_2 (Height)	.037	.001
x_3 (College aptitude)	.926	.857
Variance accounted for	1.715	1.715
Proportion of total variance	.572	.572
Proportion of common variance	1.000	1.000

c. Principal components

Variable	Component 1	Component 2	Component 3	Communality
x_1 (IQ)	.974	−.040	.221	1.000
x_2 (Height)	.073	+.997	.001	1.000
x_3 (College aptitude)	.975	−.035	−.221	1.000
Variance accounted for	1.905	.997	.098	3.000
Proportion of total variance	.635	.332	.033	1.000

d. Principal component solution, varimax rotation

Variable	Factor I	Factor II	Communality
x_1 (IQ)	.975	.016	.951
x_2 (Height)	.017	.999	.998
x_3 (College aptitude)	.975	.019	.951
Variance accounted for	1.902	.999	2.900
Proportion of total variance	.635	.332	.967
Proportion of common variance	.657	.343	1.000

*Squared multiple R, which is replaced by 1.000 for the principal-component solution in part *c*.

Table 5.14 *Three inter-related variables: two analyses*

	a. Correlation matrix		
	IQ	*Vocabulary*	*College aptitude*
IQ	.822*	.851	.902
Vocabulary	.851	.815*	.898
College aptitude	.902	.898	.875*

b. Principal-factor solution		
Variable	*Factor I*	*Communality*
x_1 (IQ)	.921	.848
x_2 (Vocabulary)	.918	.843
x_3 (College aptitude)	.956	.914
Variance accounted for	2.605	$2.605 = H^2$
Proportion of total variance	.868	.868
Proportion of common variance	1.000	1.000

c. Principal components				
Variable	*Component 1*	*Component 2*	*Component 3*	*Communality*
x_1 (IQ)	.955	.267	.126	1.000
x_2 (Vocabulary)	.954	−.279	.112	1.000
x_3 (College aptitude)	.972	−.011	−.234	1.000
Variance accounted for	2.768	.149	.083	3.000
Proportion of total variance	.923	.049	.028	1.000

d. Principal-component solution, varimax rotation				
	Factor I	*Factor II*	*Factor III*	*Communality*
x_1 (IQ)	.810	.428	.400	1.000
x_2 (Vocabulary)	.424	.814	.396	1.000
x_3 (College aptitude)	.500	.497	.709	1.000
Variance accounted for	1.086	1.094	.820	3.000
Proportion total variance	.362	.365	.273	1.000

*Square multiple R, which is replaced by 1.000 for the principal-component solution in part c.

Table 5.13b shows the principal-factor solutions for the three-variable correlation matrix with two related variables, IQ and College Aptitude, and a third variable, Height, unrelated to either of them. The initial communality (estimated from totaling the multiple R's) is 1.629 or 54.3 percent of the total variance. The first factor accounts for close to this amount of common variance. The factoring was stopped after one factor, since slightly more than this proportion of total variance was accounted for by the first factor. Note that Factor I "loads" (that is, has r_{jI} greater than .333) on the two variables that are highly correlated in the correlation matrix. This is just what we would intuitively expect: that the first and second variables define a factor we can call something like "intellectual competence" or "academic ability."

Table 5.14 shows the situation in which all variables are correlated to all others, which suggests a single general factor. Indeed the first factor is defined by all three variables and accounts for 92.3 percent of the total variance, again more than the initial estimate (from the total of the three multiple R's of 83.4 percent). Indeed, the correlation matrix is a one-factor matrix; the name of this factor might be something like "verbal skills" or "academic ability" or, possibly, "test-taking ability." Notice that the last name suggested is quite different from the first two. The choice of name for a factor always depends on considerations outside the original correlations or the factor matrix themselves.

The third approach to the problem of communality and number of factors essentially bypasses it. This is the *principal-component* method. A principal-component analysis is mathematically identical to a principal-factor solution performed on a correlation matrix with 1's in the diagonal cells. The placing of 1's in the diagonals represents a decision to analyze all of the variance instead of the common variance alone. Notice that the sum of the diagonals will now be k, which is the value of the total variance for k standardized variables. Also, of course, $r_{jj} = 1.000$ when common, specific, and error portions of each individual's score z_{ij} is included in the correlation formula.

Accounting for all the variance in k variables requires k principal components. The analysis produces them in order of decreasing proportions (accounted for) of total variance. After the fact, those components that have no loadings greater than .333, and which account for essentially none of the variance, can be dropped for interpretive purposes. Thus, if an author states that a principal-component analysis was used, but exhibits fewer than k components or factors, it is because the remaining components had essentially no loadings.

Common variance can be estimated after the fact by noting the proportion of total variance accounted for by just those components with at least one loading greater than .333. Such components are usually considered in the same way as common factors for interpretation. Since the principal-component analysis makes no use of prior communality estimates, it is technically not factor analysis. However, there is no practical difference from a reader's point of view. The methods are used in the same situations; the results of both can be interpreted according to the suggestions in section 5.5.1.

Tables 5.12c, 5.13c, and 5.14c show the principal components found for the same tables of correlations that were previously discussed in connection with principal factors. For the situation in which variables are uncorrelated, there was no common factor. But all three principal components have high loadings. The first component resembles a general factor, although one is hard pressed to define it. Similarly, the other two factors make little intuitive sense. Indeed, we can be fairly sure in this instance that there *is no* intuitive sense. Our previous estimate by way of multiple R's showed less than 5 percent of the variance to be common variance. Therefore, it is safe to conclude that the loadings in Table 5.12c represent specific variance and error variance.

Table 5.13b shows one common factor for a correlation matrix in which two variables correlate with each other but not with the third variable. In Table 5.13c, there are two principal components, each with at least one loading greater than .333. The first is defined by the two correlated variables and the second by the third variable. Together, these two principal components account for nearly all the common variance and the third can be considered error. Compare Tables 5.13b and 5.13c. The isolated variable Z does not help define *any* common factor, but defines one principal component all by itself. This is because principal factor solutions are stopped when all remaining to-be-accounted-for variance is specific variance or error variance. Isolated variables show up as one-variable components in principal-component analysis. The interpretation is not really affected by the principal-factor—principal-component distinction. The first factor in either Table 5.13b or 5.13c might be named "intellectual competence"; the specific factor Height would retain its name in Table 5.13c, whereas, in discussing Table 5.13b in a text, one would mention that the remaining 42.8 percent of the variance (the noncommon variance) was probably due to the isolated height variable.

Tables 5.14b and 5.14c agree well. There is only one principal component with at least one loading greater than .333 and there is one common factor. Both the common factor and the first principal component account for large portions of the analyzed variance; all three variables load high on them.

In general, the principal factor and principal component solutions give similar results as far as drawing conclusions is concerned. In the only instance, Table 5.12, for which this is not true, the correlations were not appropriate for any kind of factor analysis. Table 5.12a shows no common variance; thus a plan to account for common variance would be absurd. Table 5.12 was included here to illuminate the relationship between a pattern of correlations and the results of factor analyzing the correlations.

5.5.4 Rotated Solutions

You may encounter a reference to a factor matrix having been subjected to a rotation (a "varimax" or "quartimax" rotation, perhaps); you may be informed that this was done in order to attain something called "simple structure"—such statements refer to the geometric aspects of factor analysis, which are best understood by means of a figure such as Figure 5.1. The figure shows eight points located in two-dimensional space. The points represent the variables in a factor analysis; the axes represent the factors. The straight-line

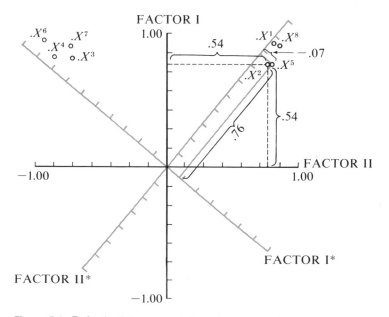

Figure 5.1 *Principal factors, with varimax rotation, for an example with eight variables.*

distance between an axis and a point corresponds to the loading of the variable on that factor. Thus, the first variable is located at point .54; .54 because it loads +.54 on both Factor I and Factor II.

The important feature of Figure 5.1 is the pair of lighter colored axes, I* and II*. They represent a defensible factor analysis of the correlations that led to the first pair of axes (factors). These other axes can be obtained by rotating the first ones about their own intersection point. The points representing variables have not been moved; their location relative to each other has not changed. What has changed in the rotation of axes is the frame of reference. Axes I and I* represent two different hypothetical characteristics as do axes II and II*. The first variable, for instance, can just as well be thought of as (a) −.07 Factor I* + .76 Factor II* as it can be thought of as (b) .54 Factor I + .54 Factor II.

In short, the positioning of the factors in space is rather arbitrary. This is true for any number, p, of factors. To reduce the arbitrariness of this procedure, sets of criteria for a good positioning have been suggested. The classical statement is that of L. L. Thurstone[7] for what he called *simple structure:* simple structure is obtained by rotating the axes until—

1. Each row of the factor matrix has one zero.
2. Each column of the factor matrix has p zeros, where p is the number of factors.
3. For each pair of factors, there are several variables for which the loading on one is virtually zero and the loading on the other is substantial.
4. If there are many factors, then for each pair of factors there are many variables for which both loadings are zero.
5. For every pair of factors, the number of variables with non-vanishing loadings on both of them is small.

These criteria simply state that the factor analysis should reduce the complexity of all the variables (items 1, 3, 4, and 5) and should do so by introducing factors that are defined in terms of only some of the variables (items 2, 5). Table 5.10b gives a principal-factor analysis after rotation. Table 5.15 has been adapted from Table 5.10b. It shows X's wherever the loadings were great ($\geq$.33) and 0's wherever the loadings were near zero ($<$.33). Indeed the solution satisfies all of the criteria of simple structure. All variables have one zero; the

[7] Thurstone, L. L., *Multiple Factor Analysis*. Chicago: University of Chicago Press, 1947.

Table 5.15   *Schematic representation of Table 5.10b*

Variable	Factor I	Factor II
x_1	0	X
x_2	0	X
x_3	X	0
x_4	X	0
x_5	0	X
x_6	X	0
x_7	X	0
x_8	0	X

columns each have four zeros; the one pair of factors has no instance of two X's (item 4 does not apply).

There are several mathematical methods of rotating the initial solution to an approximation of simple structure. The most popular is the *varimax* rotation illustrated in Tables 5.10b, 5.12d, 5.13d, and 5.14d. This method uses methods of the calculus and of matrix algebra to maximize (simultaneously for all factors) the variance of the loadings within each factor. The variance of a factor is largest when its smallest loadings tend toward zero and its largest tend toward unity. Thus, the varimax solution produces factors that are characterized by large loadings on relatively few variables.

Another commonly used analytical rotation is the *quartimax* solution, which uses mathematical methods to transform the loadings until the variance of the squared factor loadings throughout the matrix is maximized. This solution permits a general factor to emerge, whereas the *varimax* solution would not. Both the quartimax and the varimax solutions produce factors that are uncorrelated.

There exist several mathematical methods of rotation for which factors are correlated among themselves. These methods—the *quartimin*, *oblimax*, and *oblimin* techniques—are the most general because they permit, as one instance, uncorrelated factors. When an uncorrelated-factors solution is given, you should infer the investigator's judgment to have been that the dimensions of interest are uncorrelated on theoretical grounds. The use of one of the rotations permitting correlated solutions implies that the issue of correlation among factors is considered an open question.

5.5.5 Summary of Factor Analysis

The various techniques we have discussed have a common goal: reducing the complexity of the interpretation of a table of intercorrelations. Each technique attains this goal by three steps: (1) expressing each variable as a linear function of a relatively small number of hypothetical dimensions called factors; (2) expressing the known correlations in terms of the factors; (3) solving the resultant simultaneous equations. The factors are defined by (and named on the basis of) their correlations with the various original variables. Arbitrariness occurs in several places in factor analysis: in the number of common factors, in the proportion of each variable's variance that is considered common (the communality), and in whether or not the factors are correlated. Various solutions to these problems lead to the various models that exist. For the reader, the important fact is that factor analysis is an interpretive process; that is, there is no unique factor solution for a given intercorrelation table. and for a given factor matrix, there are several equally defensible interpretations; the reader's own should be one.

5.6 Other Multivariate Techniques

Correlational analysis, factor analysis, and multiple regression are the most common multivariate techniques encountered. Two others will be briefly defined. First there is *discriminant analysis,* this is a technique for assigning weights to a set of variables in such a way that the total scores will be as different as possible for two or more groups. For example, suppose we have a group of students who have been placed on academic probation in state colleges and a group of recent state college graduates. We may have scores for both groups on a battery of k tests given prior to admission. Discriminant analysis determines $\alpha, \alpha_1, \alpha_2, \alpha_3, \ldots \alpha_k$ such that the means of the score $Y = \alpha_1 X_1 + \alpha_2 X_2 + \ldots \alpha_k X_k + \alpha$ are as different as possible for the two groups. The process involves applications of the calculus and of matrix solutions to simultaneous equations. When this is done, the Y score predicted by means of the student's test scores and the alpha weights could be used as a basis for future college admissions policy.

The other common multivariate technique we shall define is *canonical correlation.* This is a technique for applying numerical weights to the variables in *two* distinct sets such that the two resulting combined scores will be maximally correlated. That is, $\alpha_1, \alpha_2 \ldots \alpha_k$ and $\gamma_1, \gamma_2, \gamma_3 \ldots \gamma_j$ are determined such that the variables $X = \alpha_1 X_1 + \alpha_2 X_2 + \cdots \alpha_k X_k + \alpha$, and $Y = \gamma_1 Y_1 + \gamma_2 Y_2 + \cdots \gamma_j Y_j + \gamma$ have a maximum common variance. The process of finding the weights requires factor analyses with two matrices.

The technique was developed to apply to situations in which two batteries of tests were administered to one group of subjects. The $X_1, X_2 \ldots X_k$ scores would represent one battery, and the $Y_1, Y_2,$

$Y_3 \ldots Y_j$ scores another. Another possible application would be for the study of pairs of persons—mothers and infants, students and teachers, twins, and so forth. In such an instance, the X scores would be for various attributes of the infant, say, and the Y scores would be for the attributes of the mother. The resulting canonical-correlation solution would give an over-all description of the presence or absence, and the patterning, of a relationship between mother variables and infant variables.

5.7 Summary

We have discussed techniques that focus on the relationships among a number (often a large number) of variables. The specific procedures vary from the inspection of a table of correlation coefficients to the computer application of advanced mathematical procedures to those same tables. The reader's most important tool is the concept of common variance and its relevance to defining a variable in terms of its correlates and to making predictions about individual scores or about one variable from knowledge of several others.

6

Nonparametric ("Distribution-free") Statistical Procedures

The adjective "distribution-free" applies to those statistical procedures that require no assumptions about the shape of the distribution of scores within the population(s) being studied. Such tests are made without assuming a specific value for the population mean(s), variance(s), or other parameters. Thus the *distribution-free* tests are also *nonparametric* tests; the latter adjective is frequently employed to describe them.

6.1 When Nonparametric Procedures are Used

Nonparametric tests avoid inferences, the validity of which depends on the actual numerical values of the raw scores. Instead, various ordering or classifying and counting procedures are applied to the scores; the resulting frequencies, ranks, or values derived from them are used as the numerical bases of inferences.

For example, suppose Marcia is assigned a score of 2 on a three-point scale of "proficiency at task." Most *parametric* tests make use of the actual value 2 and add it (to other values), square it, or perform other arithmetic operations on it; by contrast, the *nonparametric* test employs the fact that 2 is the middle score (information about *rank* or ordinal position) and information about how many subjects obtained that rank.

Investigators usually make the decision to employ nonparametric tests on the basis of either (*a*) a belief that the scores they have obtained in a particular study are such that the assumptions prevalent in common parametric tests about distributions and parameters are unlikely to be met by the scores, or (*b*) a general attitude that dictates that research conclusions should be based on as small a number of untested assumptions as can possibly be arranged, or (*c*) a determination that a particular nonparametric test is the one test most likely to reject the null hypothesis being tested if it is indeed false—that is, a determination that the nonparametric test is the most powerful test of the null hypothesis.

A small N is frequently found in studies analyzed by nonparametric procedures. In much behavioral-science research, there is no basis on which we can assume a given type of distribution for the scores. When sample size is small, investigators hesitate to make any assumptions about properties of the population distribution based on the unreliable information in the sample. Choice of a nonparametric test here is an instance of reason *a*. The reader's ability to recognize that nonparametric procedures have been used in determining the conclusions made by an author is somewhat limited. Few current journals can afford to devote space to lists of ranks or to explanations of counting procedures—or to offer any other clues. Therefore, the ability to recognize the use of these tests is primarily a matter of vocabulary—of knowing the names of the common nonparametric tests and the symbols employed in them. An inventory of these tests, their symbols, and their applications is given in Section 6.5. Before we get to that, it would be well to examine the basic rationale of nonparametric tests (Section 6.2), the concept of randomization (Section 6.3), and the controversial levels-of-measurement issue (Section 6.4).

6.2 Finite Probability. The Basis for Nonparametric Inference

The logic of nonparametric inference depends heavily on the principles of finite probability. We know that an event's probability may be defined as the long-run frequency of that event relative to other possible events; this definition is most appropriate when the number of possible events is infinite, or nearly so. But suppose the number of events is finite and countable. Consider that classic pedagogical example: the urn containing a number, N, of balls identical except for color. Suppose a number, R, of these balls are red; if a ball is blindly taken from the urn (which has been well shaken) the probability of its being red is

$$P(\text{red}) = \frac{R}{N} \qquad (6\text{-}1)$$

This conceptualization of probability indicates that the probability of an event is given by the ratio of balls constituting that event (red) to the total number of balls. The removal of a single ball is an instance of a "simple event"—a more general term. When the number of simple events is finite and countable, the definition stated by Equation 6-1 is used.

Distribution-free tests are based on the principles of finite probability apparent in the urn example: the scores are analogous to the balls in the urn; the sample size is analogous to the total number of balls. In distribution-free tests, questions are posed (a) about the probability of the ordinal positions of certain scores being as they are or (b) of a certain partitioning of the scores into groups being as it is; the null hypothesis is formulated from certain assumptions about the scores (balls) in the sample (urn). These assumptions then serve as the basis for relevant probability calculations.

Let us consider a further simple example. Suppose 10 white balls are placed in our urn and, further, suppose that five balls have been permanently imprinted with the word "fail" and five with the word "pass." Our experiment consists of shaking the urn to thoroughly mix the balls and then drawing out four of them, one by one. These four balls are then painted red (without obscuring their labels) and assigned to our control group; the remaining six balls are left white (to represent our experimental group). If our shaking was done properly and if the balls were chosen blindly, the probability of a "pass" on a red ball should be the same as the probability of a "pass" on a white ball.

This experiment is depicted in Figure 6.1. The assumption that the urn was shaken well prior to the drawing corresponds to the assumption that our subjects were assigned to conditions at random. That is, the shaking should result in giving all balls the same probability of being chosen. The notion of a blind drawing corresponds to the null hypothesis of no experimental effect; that is, under the null hypothesis, there is no systematic or purposeful selection of

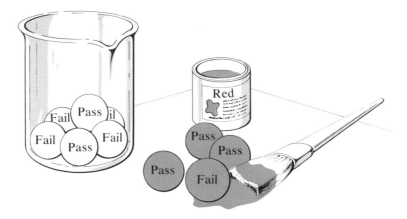

Figure 6.1 *Finite probability example.*

passes for inclusion in the group of balls to be painted red. An analogy to a systematic effect resulting from experimental manipulation would be to systematically look in our urn and *select* four "fails" to be painted red (control) and leave five "passes" and one "fail" in the control group.

There are, of course, a number of different sets of four balls that can be taken blindly from a well-shaken urn. A set of four balls that is deliberately selected from the urn is identical to one of the many four-ball sets that could have resulted from four successive blind draws. The typical nonparametric test is based on a calculation of the probability of the actual data (a certain arrangement of balls), *given* that they arose by some process analogous to a blind draw of balls from an urn.

Equation 6-1 allows us to calculate the probability of drawing a red ball on a single draw of one ball. In statistical inference, we are usually interested in more complex problems (like the previously mentioned blind drawings of four balls). For such problems, we must employ rules for calculating the probability of a complex event (four "fail" balls) from knowledge of the probabilities of the simple events (the second ball drawn was a "fail") that comprise them.

Two important rules are employed for these calculations. The first is known as the "and" rule:

> *"And" Rule:* The probability of "*A* and *B*" is given by the product of the probability of *A* and the conditional probability of *B*, *given that A* occurred.

Consider an urn containing 10 balls (five "fails" and five "passes").

Let event A be the drawing of a "fail" on the first draw and event B be the drawing of a "fail" on the second draw. Our basic definition of probability (Equation 6-1) says that

$$P(A) = \frac{5}{10}$$

since, for the first draw, there are 10 balls available, five of which are imprinted "fail." However, after the first draw has been made, there remain only nine balls in the urn, and, *given that* the first ball was a "fail," then four "fails" remain among the nine. Thus

$$P(B, \text{ given } A) = \frac{4}{9}$$

and, according to the "and" rule,

$$P(A \text{ and } B) = \frac{5}{10} \times \frac{4}{9} - \frac{20}{90} = \frac{2}{9}$$

The second important rule is known as the "or" rule and is:

"Or" Rule. If A and B are mutually exclusive events (A and B cannot both occur), then the probability of A or B occurring is given by the sum of their separate probabilities, that is,

$$P(A \text{ or } B) = P(A) + P(B)$$

In the simple 10-ball urn example, let C represent the obtaining of a "pass" on the first ball; let D represent the obtaining of two "fails" on the first two draws; let E represent the obtaining of two "passes" on the first two draws. Event D is equivalent to A **and** B and has a probability, previously calculated, of $\frac{2}{9}$. Event E consists of obtaining a "pass" on the first draw (probability $= \frac{5}{10}$) and on the second draw (given that the first ball was a "pass": probability $= \frac{4}{9}$). Thus $P(E) = \frac{2}{9}$. Now, according to the "or" rule:

$$P(D \text{ or } E) = P(D) + P(E) = \frac{2}{9} + \frac{2}{9} = \frac{4}{9}$$

Special note should be taken of one aspect of the urn example; namely, successive draws are made without replacement of previously drawn balls. This means that the size, N, of one's conceptual urn is decreasing by one on each successive draw. *Finite probability theory* is probability theory based on such drawing without replacement. It is an appropriate model for statistical procedures based on samples, since N independently selected subjects can be considered analogous to the N balls in an urn. Distribution-free procedures compare to some preset level of significance: the probability that, given all assumptions, the actual sample at hand, constituted exactly as it is, would arise. When the sample at hand has a very small probability (less than the level of significance), then the assumption of a blind draw is abandoned. An alternative notion is accepted; namely, that something systematic—no doubt the experimental manipulation—caused the sample to be as it is.

Specific examples of nonparametric hypothesis tests will be examined in the next section.

6.3 Randomization Tests for Two-Sample Cases

6.3.1 Example

Suppose we have five subjects available; two experimental treatments are to be employed, with two subjects assigned to one treatment and three to the other. Under treatment A, two subjects are asked to read a list of five nonsense syllables aloud and then to recite the list from memory. This process is repeated until a perfect recitation is obtained; the score X is the number of repetitions required for perfect recitation. Treatment B is similar, except that the three subjects read the list silently to themselves between recitations.

In this example, we will assume that our theory predicts that the utterance of the syllables aloud aids in their memorization. Thus, the hypothesis alternative to H_0 would state that, on the average, the number of repetitions required by the treatment A group of subjects would be less than the number required by the treatment B group. The null hypothesis would state that these two numbers would be equal.

Let us suppose that the numbers of trials required by our five subjects were 1, 2, 3, 4, and 6. There are 10 different ways in which these five scores can be apportioned into two groups containing, respectively, two scores and three scores. These 10 partitions are shown in Table 6.1.

Let us assume that Partition 6 corresponds to the actual data obtained; that is, the two subjects in treatment A (recite aloud) took 1 and 3 repetitions, respectively, and the three subjects in treatment B (read silently) took 2, 4, and 6 repetitions respectively. The randomization test is based on the exact probability of a difference in

Table 6.1   *Partitions of five scores into a group of two and a group of three*

	1		2		3		4		5	
	A	B	A	B	A	B	A	B	A	B
	2	1	2	1	2	1	3	1	1	3
	4	3	3	4	6	3	4	2	2	4
		6		6		4		6		6
Sum	6	10	5	11	8	8	7	9	3	13
Average	3.00	3.33	2.50	3.33	4.00	2.67	3.50	4.50	1.50	4.33

	6		7		8		9		10	
	A	B	A	B	A	B	A	B	A	B
	1	2	1	2	1	2	3	1	4	1
	3	4	4	3	6	3	6	2	6	2
		6		6		4		4		3
Sum	4	12	5	11	7	9	9	7	10	6
Average	2.00	4.00	2.50	3.67	3.50	3.00	4.50	2.33	5.00	2.00

means, $\bar{X}_B - \bar{X}_A$, as large or larger than the one observed. The 10 possible mean differences are listed in order of magnitude in Table 6.2. Of the 10 mean differences, 2 (Partitions 5 and 6) are equal to or numerically larger than the difference for Partition 6. Thus the relevant probability for the suggested one-tailed test is $\frac{2}{10}$ or 0.2. The observed difference would occur by chance 20 percent of the time, given the original 5 scores. This difference would not be significant

Table 6.2   *Statistics derived from Table 6.1: the basis of a randomization test of H_0*

Partition	$\bar{X}_A$	$\bar{X}_B$	$\bar{X}_B - \bar{X}_A$
(5)	1.50	4.33	2.83
(6)	2.00	4.00	2.50
(7)	2.50	3.67	1.17
(2)	2.50	3.67	0.84
(1)	3.00	3.33	0.33
(8)	3.50	3.00	−0.50
(4)	3.50	3.00	−1.00
(3)	4.00	2.67	−1.33
(9)	4.50	2.33	−2.17
(10)	5.00	2.00	−3.00

at the customary level of 0.05. In fact, no mean difference would be for $N = 5$. For larger N, significant differences could be obtained.

There are five things about this example that have important implications for all nonparametric procedures: (1) The calculated probabilities are based *only* on various possible partitions of the five original scores. No attention is given to other scores that might have arisen if another sample of five persons had been used; no assumption about the five scores being a random sample of any population is made; all partitions of five scores into two and three are considered, but all possible sets of five scores are not.[1] (2) There is no assumption inherent in the test about the relevance of the conclusion to some undefined population of persons. Such statements often are made following the calculation of randomization tests, but they depend for their justification on considerations outside the realm of statistics. (3) The probability calculations do not depend in any way on assumptions about the nature of the distribution of the score "number of repetitions." In Table 6.2, the rank order of partitions is the same, no matter whether mean of treatment B scores is used or the difference between means, $\bar{X}_B - \bar{X}_A$, or the sum of the B scores. Thus, the calculations may be, and usually are, based on the sum of B scores, since it is the easiest value to obtain. (5) The inference that the probability of a group difference numerically equal to or greater than the obtained difference is 0.20 depends only on the rank position of the sum; that is, the magnitude of the original scores, as such, does not enter into the logical process.

6.3.2 Example

We now consider an example that is the same as the example in Section 6.3.1 except that a two-tailed null hypothesis will be assumed. Look again at Table 6.2, which ranks the sums of the scores in treatment B. The *sum B* for Partition 9 is as many ordinal positions *away* from the middle (between Partitions 1 and 8) as the obtained *sum B* (Partition 6). Similarly, the *sum B* for Partitions 5 and 10 are equally far away from the middle. Thus, the probability of a *sum B* as deviant (from the middle) or more deviant (from the middle) than the obtained *sum B* would be equal to the sum of the probabilities for Partitions 5, 6, 9, and 10, or $1/10 + 1/10 + 1/10 + 1/10 = 4/10 = 0.4$. Note that this total is twice the probability for the same

---

[1] However, if the five scores can be regarded as a random sample of some larger population of scores, then the conclusion drawn here (do not reject H_0) can and should be generalized to that population.

data and a one-tailed hypothesis. Obviously, in this case, the null hypothesis must not be rejected at any reasonable level of significance. With a larger N, a two-tailed null hypothesis might be rejected.

6.3.3 Example

Now suppose we conduct the same memorization experiment, but that this time we have two lists and four subjects. Each subject memorizes one of the lists under treatment A (recite aloud) and the other list under treatment B (read silently). In order to eliminate considerations of specific effects of order-of-presentation and list-treatment combinations, one subject is assigned to each of the possible experimental sequences, which are as follows:

First	*Second*
List 1, treatment A	List 2, treatment B
List 1, treatment B	List 2, treatment A
List 2, treatment A	List 1, treatment B
List 2, treatment B	List 1, treatment A

The difference scores, $D = X_B - X_A$, are used for analysis. Here, X_B stands for number of repetitions under treatment B, and X_A stands for number of repetitions under treatment A. D, the algebraic difference, is positive whenever X_B exceeds X_A, and negative whenever X_A exceeds X_B.

Table 6.3a *All possible arrangements of plus and minus signs for four* $D = X_B - X_A$ *scores*

	1	2	3	4	5	6	7	8
	-2	-2	-2	-2	$+2$	-2	-2	$+2$
	-1	-1	-1	$+1$	$+1$	-1	$+1$	-1
	-3	-3	$+3$	$+3$	$+3$	$+3$	$+3$	$+3$
	-1	$+1$	$+1$	$+1$	$+1$	-1	-1	$+1$
Sum of $+D$	0	1	4	5	7	3	4	6

	9	10	11	12	13	14	15	16
	$+2$	-2	$+2$	$+2$	-2	$+2$	$+2$	$+2$
	-1	$+1$	$+1$	$+1$	$+1$	-1	-1	$+1$
	$+3$	-3	-3	-3	-3	-3	-3	$+3$
	-1	-1	-1	$+1$	$+1$	-1	$+1$	-1
Sum of $+D$	5	1	3	4	2	2	3	6

Table 6.3b *The "sum of $+D$" statistics listed from left to right in descending rank order*

7	6	6	5	5	4	4	4	3	3	3	2	2	1	1	0

Suppose our four subjects provided D scores of 2, 1, 3, and -1. Table 6.3 presents *all possible* arrangements of plus and minus signs for the four difference scores, together with the sum of positive D's that would result for each arrangement. Arrangement 16 corresponds to the data at hand, and its "sum of $+D$" is 6, one of the six extreme scores. (The extreme scores are 7, 6, 6 on the high end of the list in Table 6.3b, and 0, 1, 1 on the low end. Thus, for a two-tailed test, the obtained sum of $+D$ has a probability of $\frac{6}{16}$, or 0.375, which exceeds customary α. Therefore, a two-tailed null hypothesis would not be rejected.

If we predicted, as before, that treatment A would be more effective than treatment B in facilitating learning, a one-tailed test would be used. The alternative (research) hypothesis would state that $D = X_B - X_A$ is positive; that is, that more repetitions would be needed under treatment B than under treatment A. The number of arrangements shown in Table 6.3b to have a sum of $+D$ as large or larger (in *one* direction) than the obtained value of 6 is 3. Thus, the probability of $D \geq 6$ is $\frac{3}{16}$ or 0.188. The one-tailed test permits of rejection of H_0 only at the 0.188 level of significance.

6.4 The Levels of Measurement Issue

It has been argued[2] that nonparametric tests must be chosen whenever scores are based on measurement that is not on at least an interval scale; this argument rests on the assertion that there should be an exact correspondence between the structure of a set of scores and the structure of the arithmetic employed in calculating statistics for inferential use. Actually, however, arithmetic calculations are independent of the ultimate meanings of the numbers (scores) used to make them. Furthermore, statistical inference depends on assumptions about the nature of distributions of scores for groups of individuals and is independent of the kinds of measurement procedures that led to the scores in the first place.

[2]See, for example, Siegel, S., *Nonparametric Statistics for the Behavioral Sciences*, New York, McGraw-Hill, 1956.

Even though the issue is not relevant to the investigator's choice between nonparametric or parametric tests for his analyses, the reader of the literature does need to be concerned about levels of measurement and about the relationship between the measurement procedures employed to obtain the scores and the investigator's verbal interpretations of his results. Most often, these verbal interpretations occur in the "discussion" section of the article, wherein the investigator attempts to integrate his findings with those of others and with theory. Tables 6.4a, 6.4b, and 6.4c present two alternative analyses for a set of nine scores. The analysis-of-variance procedure of Table 6.4a might be chosen by an investigator if he feels that the required assumption were not violated severely: namely that the three populations from which the three samples came had normal distributions and equal variances. The procedure based on ranks would be chosen if the investigator believed these assumptions to be untenable. His decision at this point does not depend on the particular measurement procedures he employed to obtain the nine scores nor on any other rationale he may have for giving them meaning.

Both statistical tests lead to a rejection of the null hypothesis of equal means (Tables 6.4a, 6.4b) or medians (Table 6.4c). Thus, in either case, the investigator would concentrate, in his discussion section, on giving meaning to the differences.

The careful reader will compare the statements made by the investigator with the level at which measurements were obtained. When data are in categories such as "lower class," "lower-middle class," and "middle class," or "male" and "female," then conclusions should be expressed in terms of "same" and "different." (For example, "There was a sex difference, with boys passing more items than girls." is a proper conclusion based on categorical data.) When data are ranks or other ordinal measures, then conclusions should be expressed in terms of "greater than," "less than," and "the order of treatments

Table 6.4a *Data from a hypothetical experiment with three treatments and nine subjects*

	Treatment 1	Treatment 2	Treatment 3
	2	3	4
	0	2	4
	0	1	3
Means	0.67	2.00	3.67
Medians	0	2	4

was . . ." (For example, "The experimental group subjects performed better than the control group subjects." is a proper statement based on ordinal data.) When data are truly on an interval scale, precise statements may be made. (For example, "Treatment Groups I and

Table 6.4b *Calculations for analysis of variance of the data in Table 6.4a*

	Treatment 1	Treatment 2	Treatment 3	Total
$\sum X$	2	6	11	19
$\sum X^2$	4	14	41	59
n_j	$n_1 = 3$	$n_2 = 3$	$n_3 = 3$	$N = 9$

1. Total sum of squares

$$SS_T = \sum_{i=1}^{3} \sum_{j=1}^{3} X_{ij}^2 - \frac{\left(\sum_{i=1}^{3} \sum_{j=1}^{3} X_{ij}\right)^2}{N} = (2^2 + 0^2 + 0^2) + (3^2 + 2^2 + 1^2) + (4^2 + 4^2 + 3^2)$$
$$- 19^2/9 = 59 - 40.11 = 19.89$$

2. "Between" sum of squares

$$SS_B = \frac{1}{3} \sum_{j=1}^{3} \left(\sum_{i=1}^{3} X_{ij}\right)^2 - \frac{\left(\sum_{i=1}^{3} \sum_{j=1}^{3} X_{ij}\right)^2}{N} = \frac{1}{3}(2^2 + 6^2 + 11^2) - 40.11 = 53.67 - 40.11$$
$$= 13.56$$

3. "Within" sum of squares

$$SS_W = \sum_{i=1}^{3} \sum_{j=1}^{3} \left(X_{ij} - \frac{\sum_{i=1}^{3} X_{ij}}{n_j}\right)^2 = (2 - .67)^2 + (0 - .67)^2 + (0 - .67)^2 + (3 - 2.00)^2$$
$$+ (2 - 2.00)^2 + (1 - 2.00)^2 + (4 - 3.67)^2$$
$$+ (4 - 3.67)^2 + (3 - 3.67)^2 = 5.33$$

Table 6.4c *Summary table for analysis of variance*

Source	df	Sum of squares	Mean square	f
"Between" treatments	2	13.56	6.780	7.618*
"Within" treatments	6	5.33	.889	
Total	8			

*$p < .025$

Table 6.4d   *Calculations for a nonparametric test of H_0 for the data of Table 6.4a*

1. First the scores are converted to ranks.

Rank order	Scores	Assigned rank*
1	0	1.5
2	0	1.5
3	1	3
4	2	4.5
5	2	4.5
6	3	6.5
7	3	6.5
8	4	8.5
9	4	8.5

2. Then Table 6.4a is rewritten with the entries being ranks instead of raw scores. For each score the appropriate rank is selected from the table in step 1.

	Treatment 1	Treatment 2	Treatment 3
	4.5	6.5	8.5
	1.5	4.5	8.5
	1.5	3.0	6.5
Sum of ranks	7.5	14.0	23.5

3. Then the three sums of ranks for the three treatment groups are compared. Under the null hypothesis (of no treatment difference), the nine ranks would be distributed among the three treatment groups at random. Thus, we would expect the sums of ranks to be approximately equal if H_0 is true. If the sums of ranks are equal, they have zero variance.

Thus, H, the variance of the sums of ranks is used as a test statistic.

In this case $H = 27.31$ and when all possible values of H for three groups of three are considered, the probability of a value this large or larger is less than .01.

H_0 is therefore rejected.

The procedure just illustrated is known as the Kruskal-Wallis one-way analysis of variance by ranks and is further discussed in Section 6.5.

---

* When two scores are numerically equal the two ranks they would otherwise occupy are averaged and this average rank is assigned to both scores.

II differed by approximately twice the amount of difference between Treatment Groups II and III.")

It is rare in the behavioral sciences to have measures on an interval scale with regard to the variable of ultimate interest, though an interval-scale measure of some intermediate variable may be available. For example, suppose scores in Tables 6.4a, 6.4b, and 6.4c represent "trials to criterion," that is, the number of repetitions of a list of words before a subject can recall the list from memory. In this example, the scores represent the intermediate variable, and they may indeed be an interval measure of something like "elapsed time" or "effort."

Most likely, an investigator using the "trials to criterion" scores is attempting to study a dimension like "ease of learning." The treatments may represent different properties of the lists that are theoretically supposed to make them difficult or easy to learn. It is not at all clear that equal amounts of learning occur on each trial; indeed, most theories of learning would hold that assumption untenable. Suppose a subject learns eight items of a 10-word list on trial one, and a ninth on trial two. On trial three he picks up the tenth but forgets one he knew, resulting in a net score of 4 correct. One more trial is required before a perfect recitation is obtained; thus our subject has a total trials-to-criterion score of 4. However, this is probably not an interval measure of how difficult or easy to learn the list was. It is doubtful that the difference between the score "4 trials-to-criterion" and zero is the same as the difference between 8 trials and 4 trials. To reiterate, statements about the magnitude in trials-to-criterion of differences between treatment groups should not be automatically translated into assertions about the relation in "ease of learning" of different lists for the individuals in the treatment groups.

It is important to keep in mind the distinction between measurement and statistical inference. *Measurement* is a process whereby scores are assigned to individuals, whereas *statistical inference* is a process that deals with groups of scores. The issue of levels of measurement is relevant to the relationship between the measurement process and the ultimate verbal conclusions or generalizations an investigator may wish to make about his data. Statistical tests are used by him as an intermediate step to answer such questions as, "Do the means of my groups differ?" or "Are the two sets of scores I obtained related?" Thus, while the levels-of-measurement issue is often discussed in the context of statistics, it is entirely separate from and independent of the choice between parametric and nonparametric procedures for statistical inference.

6.5 An Inventory of Nonparametric Tests

When you are confronted with the results of a nonparametric hypothesis test or estimation procedure, what you need most is a kind of annotated glossary that describes each test and tells something about how to read the pertinent tables and interpret the pertinent symbols. Such a glossary is presented in this section: the most commonly encountered nonparametric tests are presented in Sections 6.5.1 through 6.5.3, with tests arranged alphabetically within sections. Several nonparametric correlation coefficients will be presented in Section 6.5.4.

6.5.1 Tests for a Single Group

6.5.1.1 The binomial test. This test is appropriate when all members of a sample are classified into one of two mutually exclusive and exhaustive (that is, complementary) categories such as "male" and female." The subjects might be volunteers for a study of hypnotic susseptibility, and the null hypothesis might be that such a study is equally attractive to men and women. The binomial probability distribution gives the probabilities with which different proportions of men to women would appear in samples of fixed size, if some *a priori* hypothesis about p, the "true" proportion of, say, men, is true. More generally, the binomial probability distribution is the distribution of the number, X, of observations in n independent trials that fall into one of two possible complementary categories, which have probabilities of occurrence on any given that of p and $1 - p$ respectively. It can be shown that the probability that X is equal to some value, k, is given by the formula

$$p(X = k) = \frac{n!}{k!(n - k)!}(p)^k(1 - p)^{n-k} \qquad (6\text{-}2)$$

where p is the probability of the category on any one trial.

Suppose ten volunteers show up for the hypnosis study and eight are women. The binomial distribution for $p = .5$ and $n = 10$ is used to derive the probability of eight *or more* women turning up in a sample of ten, given the null hypothesis. This probability is equal to the sum of the values of Equation 6-2 for $k = 8, 9$, and 10, or 0.055. The test would most likely be reported in the text of the results section of a report. A typical statement might read, "Ten subjects volunteered, and eight of these were women. Since this high a proportion of women would occur by chance less than 5.5 times in 100, we concluded that women volunteer to be hypnotized more readily than men." A specific reference to the *binomial* distribution is often omitted from such statements, so the reader must learn to identify "binomial" with (1) two complementary categories, (2) one sample, (3) some *a priori* hypothesis about the proportions p and $1 - p$. This hypothesis is often summarized as "chance," that is, $p = 0.50$.

6.5.1.2 The chi-square test for one sample. The family of chi-square tests are all based on the chi-square probability distribution from which theoretical statisticians calculate probabilities for various values of the statistic, χ^2, defined as

$$\chi^2 = \sum z^2 = z_1^2 + z_2^2 + z_3^2 + \cdots + z_k^2$$

where the K z's (1) are standard scores, (2) have normal distributions, and (3) are independent. The applications are based on the fact that the difference between the observed frequency for some category and that expected on the basis of the null hypothesis, $D = O - E$, is normally distributed for large enough samples. The statistic used in most applied situations is

$$\chi^2 = \sum_{j=1}^{k} \frac{(O_j - E_j)^2}{E_j} \tag{6-3}$$

The chi-square test for one sample is used when the apportionment of the N sample subjects into k mutually exclusive and exhaustive categories is to be compared with that expected on the basis of theoretical considerations. The categories can represent the classes of a qualitative variable or they can represent the partitions of a discrete or continuous variable. Since the fit of observed to theoretical apportionment is being studied, this application is often referred to as a goodness-of-fit test. Suppose, for example, that a four-alternative multiple-choice opinion test is administered to 100 subjects. The results might appear as in Table 6.5. The Expected Frequency column lists 25, 25, 25, and 25 as the values expected if the alternatives are equally likely. The precise null hypothesis would be stated in terms of $p_1, p_2, p_3, \ldots p_k$, the expected proportions. Thus, in this example, H_0 is:

$$p_1 = p_2 = p_3 = p_4 = 0.25$$

A precise statement of this null hypothesis may be omitted, and it is instead implied by the expected frequency column.

The chi-square value listed at the lower-right corner of the table

Table 6.5 *Theoretical and observed frequencies of four responses to a multiple-choice opinion item (Question: "The best response to disobedience in a small child is to . . .")*

Answer	Frequency	Expected Frequency
"Ignore it."	15	25
"Spank the child."	15	25
"Explain the reason for request."	35	25
"Bawl him out."	35	25
TOTAL	100	100

$$x^2 = 16.00 \qquad df = 3 \qquad p < .01$$

is calculated according to Equation 6-3, and the p value indicates the probability of a χ^2 at least as large as the one obtained. The symbol, *df*, refers to degrees of freedom and is always equal to the number of the terms in Equation 6-1 that are statistically independent. This number is always $k - 1$ for one sample test. The expected value of χ^2 is equal to $k - 1$, the degrees of freedom; thus, the more categories there are, the higher the value of χ^2 required for significance.

    6.5.1.3 The Kolmogorov-Smirnov test for one sample. This test is a goodness-of-fit test for observed and expected frequencies when the underlying variable is continuous or discrete. When the underlying variable is continuous, the test is exact; when it is discrete, the test is approximate. However, unlike the chi-square test, the Kolmogorov-Smirnov test is often applied by behavioral scientists to *ordinal* scores. Thus, it might be used with data like those in Table 6.6, in which the scores represent the "developmental level" of each

Table 6.6 *Hypothetical choices among four alternative solutions to a moral dilemma*

Statistic	Developmental level of solution			
	I	II	III	IV
Frequency	2	9	27	13
Cumulative frequency	2	11	38	40
Theoretical cumulative frequency	10	20	30	40

$$D = \frac{|38 - 30|}{N} = \frac{8}{40} = .2 \qquad p < .10$$

of several solutions to a hypothetical moral dilemma. The test is appropriate if a null hypothesis is chosen that proposes values for the cumulative frequencies represented in the second row; the null hypothesis might well predict that all alternative solutions to the dilemma are equally likely, whereas the alternative hypothesis might merely state that they have different likelihoods.

The test statistic D is $1/N$ times the absolute value of the largest difference between observed and expected cumulative frequency. The sampling distribution of D has been determined, and critical values of D have been tabulated and are available to investigators.[3] It would not be unusual for an investigator to report only the name of this test and the p value associated with the performance of it with his data. Thus, the information in Table 6.6 might well appear in print in the form of a paragraph like this:

"Three-quarters of the subjects chose the solution at Level III, the dominant developmental level in their age group. Only two subjects chose the Level I or Level II stories, and nine chose the Level IV story. The Kolmogorov-Smirnov test was employed to test the hypothesis of equally likely alternatives and that hypothesis was rejected at the 0.10 level of significance."

6.5.1.4 The runs test for one sample. This test is similar to the randomization tests illustrated in Section 6.3. It is useful for answering the question: "Is a single sample a random one with respect to some dichotomous trait?" Suppose the average reading-test score of a city's first-grade population is 50. An investigator might wish to establish the fact that a given sample of 20 first graders, composed of 10 above-average readers and 10 below-average readers, was chosen strictly at random from the population. The test of randomness is performed by examining the sequence in which the children were chosen. The children's names (or other ID) are listed in order, from first chosen to last chosen; in this example, 1 to 20 as in Table 6.7. Next to each name is an A or a B, standing for above average and below average, respectively. The number of runs in which there is a succession of A's or B's is calculated, and the probability of a number so small is found from the probability distribution for the number of runs in a sequence of size N, containing n_A A's and n_B B's. This distribution is, in turn, determined from a listing of all possible sequences of A's and B's. The results of a runs test with the data in Table 6.8 might be reported in a single sentence: "Each

---

[3]Tables for this and the other nonparametric tests discussed in this chapter are available in Siegel, S., *Nonparametric Statistics for the Behavioral Sciences*, New York, McGraw-Hill, 1956, Appendix.

Table 6.7 *Data for a "runs" test on a sample of 20 children*

Sequence	Child	Reading-test status
1	Joe	A
2	Mary	B
3	Jim	A
4	Bill	A
5	Beth	A
6	Linda	B
7	Ray	B
8	John	A
9	Bob	B
10	Jane	A
11	Elaine	B
12	Tom	B
13	Karen	B
14	Carolyn	B
15	Jack	A
16	Helen	A
17	Judy	A
18	Ralph	R
19	Eric	A
20	Hilda	B

Number of runs = 12 $p < .05$

child in the sample of 20 was classified according to their above- or below-average status on the reading test; a runs test was applied and revealed that the sample, as constituted, could well be a random sample from the city's first-grade population with respect to the reading score." Notice that such a statement is made on the basis of an acceptance of the null hypothesis. That is, the statistical model is set up so randomness is the null hypothesis; therefore, randomness is a property of a sample that can be accepted but not proved.

6.5.2 *Tests for Two Samples*

The simplest and, in many cases, the most powerful statistical tests for comparing two groups are the randomization tests illustrated in Section 6.4. A number of other tests will be discussed in this section. Each of these is appropriate to either two independent groups or to the situation wherein there are two (repeated) measures for each member of one group.

6.5.2.1 The chi-square test for r × C contingency tables. This test is appropriate when individuals have been simultaneously classified on two variables, each of which is defined by two or more mutually exclusive and exhaustive categories. The categories can represent the classes of a qualitative variable or they can represent the partitions of a discrete or continuous variable. An example of results from such a classification appears as Table 6.8a, which is called a *contingency table*. The null hypothesis for Table 6.8a (or any similar table) is that the proportions of individuals in the various Expected Education categories is the same for the two sexes.

In the computation of chi square, the marginal proportions are used in estimating the expected frequencies required by Equation

Table 6.8a *Educational aspirations of 100 hypothetical twelfth graders: a contingency table to illustrate chi-square test for independent groups*

Expected education	Sex		Total
	Girls	Boys	
Advanced degree	1	7	8
Four-year college degree	10	25	35
Junior college graduation	9	10	19
High school graduation	30	8	38
TOTAL	50	50	100

Contingency coefficient = 0.474 $\chi^2 = 23.71$ $df = 3$ $p < 0.001$

Table 6.8b *Proportions based on the entries in Table 6.8a*

Expected Education	Sex		Over-all
	Girls	Boys	
Advanced degree	0.02	0.14	0.08
Four-year college degree	0.20	0.50	0.35
Junior college graduation	0.18	0.20	0.19
High school graduation	0.60	0.16	0.38
Over-all	.50	.50	

6-3. Thus, the expected frequency under H_0 in the upper-left cell is "4," found by multiplying the marginal proportion for "advanced degree," 8/100, by the number, 50, of girls.

The chi square shown was found by entering the observed frequencies from Table 6.8, together with their corresponding expected frequencies, into Equation 7-1. As with any chi square, the probability depends on the degrees of freedom; for this table there are three degrees of freedom. In general, for a contingency table consisting of r rows and c columns, there are $(r - 1)(c - 1)$ df's where r is the number of rows and c is the number of columns.

When chi square is large enough to warrant rejection of H_0, the reader will want to examine the contingency table to see where the largest deviations from expected values occur. This is best done by considering the frequencies as proportions of the total N for their group. Thus, Table 6.8b of proportions has been prepared from Table 6.8a by dividing each frequency in Table 6.8a by 50, the number of girls (one independent group) and also the number of boys (the other independent group). As we look at Table 6.8b, we see at once that large discrepancies in proportions occur for the "Four-year college degree" and "High school graduation" rows. There is no discrepancy to speak of in the "Junior college" row, and a moderate one in the "Advanced degree" row. It is tempting to summarize the results by saying that the educational aspirations of boys are higher than those of girls, but this statement would not be entirely accurate. To be specific, it would not apply to the nine girls and 10 boys who aspire to complete junior college. Since junior college is above the modal aspiration of girls and below that of boys, the choice may represent high aspiration in girls and low aspiration in boys. Further information would be needed for elaboration of this statement. The important points to remember follow. (1) Look at the patterns of proportions in the two groups. (2) A general statement of trend does not apply to all individuals; modify it accordingly. (3) All of this *post hoc* analysis is made without reference to the sampling distributions for the various relevant proportions—hence, caution is required.

The best reader-check on interpretations of the patterns of proportions is the question, "Does the study contain other data that support this way of looking at the results?" If not, the conservative statement that "the groups differ" will have to satisfy our desire for firm conclusions; verbal analyses can of course be offered as interesting speculations.

6.5.2.2 The median test for two independent groups. This test is a commonly used modification of the chi-square test for two groups. In this instance, the classification used is derived from a continuous or discrete quantitative variable, X, for which the assumptions of the parametric tests cannot be reasonably assumed to apply. The median of X for the two groups combined is determined. Then, two categories are defined: above the median and "at or below" the median. A two-by-two contingency table may be formed on the basis of classification by group membership on the columns and above or below the combined median on the rows. A chi-square statistic with one degree of freedom is then calculated. If the null hypothesis is rejected, then the conclusion is that the groups differ with respect to median score (two failed) or that one of the group's medians exceeds that of the other group. The probability values associated with values of chi square are, as stated elsewhere in the book, inherently two-tailed. The one-tailed alternative hypothesis is accepted on the basis of logical considerations outside the realm of statistics. An illustration of the median test procedures is provided in Tables 6.9a and 6.9b.

6.5.2.3 The Fisher exact probability test. This test is appropriate for 2×2 tables for which χ^2 cannot be used. In fact, the chi-square test is an approximation to the Fisher test. The appropriate use of the χ^2 approximation depends on the approximate normality of the distribution of $(O - E)$ deviations. When an expected frequency is

Table 6.9a *Hypothetical performance ratings for two groups*

Experimental group ($N_1 = 16$)	Control group ($N_2 = 16$)
5	5
5	3
5	2
5	2
4	2
4	2
4	2
4	2
3	2
3	1
3	1
3	1
3	1
3	1
2	0
1	0

Combined median (32 cases) = 2

Table 6.9b   *Data of Table 6.9a recast for median test*

Location of score	Experimental group	Control group	Total
Above median	15	1	16
Median or below	1	15	16
	16	16	32

$$\chi^2 = 31.87 \qquad df = 1 \qquad p < 0.001$$

less than 5, this assumption becomes untenable. The small expected frequency can arise (*a*) because the total *N* is small or (*b*) because the number of persons in one of the categories is small. In either event, the Fisher exact-probability test is substituted for the chi-square approximation.

The Fisher exact-probability test is based on the hypogeometric-probability distribution. The reader need not concern himself with this distribution except to note that its formula gives probabilities for various apportionments of the total frequency in a 2×2 table when the row and column totals (often called "marginals") are regarded as fixed. There exist convenient tables that list the probabilities of any given apportionment *or one more extreme in the same direction*, that is, one with even more of the total frequency apportioned to the two high frequency boxes. These tables are usually consulted in testing hypotheses. When two-tailed alternative hypotheses are involved, the tabled probability for extreme apportionments in the opposite direction must be added to the previously determined probability. In Table 6.10a presents all possible apportionments of frequency for an *N* of 6 and marginal frequencies of 3. Suppose part *a* of Table 6.10a corresponded the the obtained data. The probability of data apportioned in the manner of this table, where the marginal frequencies are all 3, may be shown to be 0.05. The obtained *p* happens to equal a commonly acceptable level of significance, and the null hypothesis of no difference would be rejected. Table 6.10b gives an example with $N = 15$ for which the probability is greater than 0.05. The interpretation of the exact probability test of part *a* of Table 6.10a may be expressed in the following sentence: "Groups I and II differed, with Group I having a significantly higher proportion of males than Group II. This difference was significant at the 0.05 level for a one-tailed test.

6.5.2.4 McNemar's chi-square test for the situation of two repeated measures. Often a contingency table results from the joint classification of persons into categories on two different variables. For

Table 6.10a   *All possible apportionments of $N = 6$ into the four cells of a 2 by 2 table with all marginals $= 3$*

	(a)			(b)	
	Group I	Group II		Group I	Group II
Male	3	0	3	2	1	3
Female	0	3	3	1	2	3
	3	3	6	3	3	7

	(c)			(d)	
	Group I	Group II		Group I	Group II
Male	1	2	3	0	3	3
Female	2	1	3	3	0	3
	3	3	6	3	3	6

Table 6.10b   *Data for a Fisher exact probability test*

	Group I	Group II	
Male	5	3	8
Female	0	7	7
	5	10	15

Fisher $p < .05$ Phi coefficient* $= .67$
Corrected phi coefficient* $= 1.00$

*See Section 6.5.4.5 for discussion of these coefficients.

example, the classifications may be responses to an attitude item on two different occasions—perhaps before and after an attempt at persuasion. For such a table, a chi-square statistic may be calculated. This calculation does not proceed *directly* from Equation 6-3. Instead, the null hypothesis of "no change from before to after" is tested by considering only those individuals who changed. In Table 6.11 this would be those individuals in the upper-left and the lower-right cells. In other words, under H_0, if change occurs at all, it is as likely to be in one direction as it is to be in the other. Chi square is calculated from Equation 6-3 with this modification. Only two cells are used; for both of them, the expected frequencies are one-half of the total of the observed frequencies in the two cells. The resultant χ^2 has one degree of freedom and is interpreted in the usual way; that is, χ^2 statistics large enough to have probability less than 0.05 or 0.01 authorize the conclusion that significant change occurred.

Table 6.11 *Hypothetical data from an attitude-change study*

		After persuasive movie		
		Con	Pro	Total
Before persuasive movie	Pro	1	10	11
	Con	5	13	18
	Total	6	23	29

$\chi^2 = 10.29$ $df = 1$ $p < .001$ Phi coefficient* $= -.22$
Corrected phi coefficient* $= -.55$

*See Section 6.5.4.5 for discussion of these coefficients.

6.5.2.5 The Mann-Whitney U-test. This test is used to test the significance of differences in central tendency between independent groups when the scores are ranks or when ranks have been substituted for the original scores. In this latter case, the test becomes relatively efficient and competes with the t test for use in these situations. The scores from both groups are ranked from 1 through $n_1 + n_2 = N$. Then a statistic, U, is calculated that gives the number of times in the grand ranking that a subject from one group preceded a subject from the other group. For example, in the data in Table 6.12, U is calculated as $1 + 2 + 1 = 4$, the number of times that a Drug subject was preceded in the grand ranking by a Placebo subject. The calculation of U counts differences in one direction and thus adjustments must be made where using tables of U for a two-tailed

Table 6.12 *Ranking of individuals from two groups according to their reaction times to a visual stimulus*

Rank	Group
1	Drug
2	Placebo
3	Drug
4	Drug
5	Placebo
6	Drug
7	Placebo

Mann-Whitney $U = 4$ $p < .10$

test. The sampling distribution of U under H_0 is known and tables are available. The investigator using a Mann-Whitney test usually will cite only the magnitude of U and the level of significance it attained. Very often the scores used in a Mann-Whitney test are seemingly ordinal in nature but are not ranks themselves. However, the test actually is performed by the substitution of ranks. In any case, the investigator will report median scores for his two groups, and then he will state that a Mann-Whitney test was employed and the obtained U was such and such. This is perfectly appropriate; it is to be assumed that the scores were ranked (with ties handled by appropriate means) and then the calculation of U performed. It should be noted that the Mann-Whitney test is equivalent to the Wilcoxon Test for two independent samples.

 6.5.2.6 The Kolmogorov-Smirnov two-sample test. This test is used to test the similarity of two independent frequency distributions— those for, say, an experimental group and a control group. It is appropriate when the scores are taken from a discrete or continuous distribution and when the assumptions for the t test are not assumed to be met. An example of appropriate data appear in Table 6.13. The test uses the maximum difference

$$D_i = \frac{\text{cum.}f_{Ii}}{n_I} - \frac{\text{cum.}f_{IIi}}{n_{II}} \tag{6-4}$$

where $\text{cum.}f_{Ii}$ is the cumulative frequency up to and including interval i in distribution I, where $\text{cum}f_{IIi}$ is the cumulative frequency up to that same interval as distribution II, where n_I is the total frequency in distribution I, and n_{II} is the total frequency in distribu-

Table 6.13 *Hypothetical IQ data for two groups*

Score	Classroom I		Classroom II		D
	Frequency	Cumulative frequency	Frequency	Cumulative frequency	
150 and above	1	36	1	35	1
140–149	1	35	1	34	1
130–139	2	34	1	33	1
110–119	11	29	13	28	1
100–109	13	18	9	15	3
90–99	3	5	5	6	−1
80–89	2	2	1	1	1
	$|D_{\text{max.}}| = 3$		$p > 0.10$; not significant		

tion II. This is the maximum difference between the cumulative relative-frequency distributions. For a one-tailed test, only differences in the predicted direction are examined for maximum, whereas for a two-tailed test the maximum absolute value of D is employed.

The sampling distribution of D is known and tables of it are consulted for stating the significance of a given D. The Kolmogorov-Smirnov two-sample test would be elected in preference to the Mann-Whitney U test in cases in which there are many ties. (Whenever several individuals receive the same score, we have ties.)

6.5.2.7 The Wald-Wolfowitz runs test. This test is appropriate for testing differences between two independent groups when (*a*) the alternative (to H_0) hypothesis is that the groups differ in some way—variance, skewness, or central tendency, and (*b*) the score represents a point on some underlying continuum—for example, reaction time. The test is performed by ranking the $N = n_1 + n_2$ scores in a single series and then calculating the number, 4, of runs of subjects of one group which results. The statistic, r, will be small if the groups differ in variability or if they differ in median. Table 6.14a and Table 6.14b show two sets of data to illustrate the Wald-Wolfowitz test. Table 6.14a shows a situation in which the two group distributions have the same median but different variances. In Table 6.14b the two groups have the same variances but different medians. In both cases, r is small relative to a possible r of 10.

The exact probabilities of r for given sizes of n_1 and n_2 can be determined by application of elementary probability theory. The investigators will usually consult a table of critical values of r and will report a significance level reached. Since a small r may be associated with one or all of several different kinds of group difference, the interpretation of a Wald-Wolfowitz test requires some additional comment about the nature of the group differences. For small N, this can come in the form of a table such as Table 6.14; otherwise, interval-frequency distributions or graphs of them are helpful. In the rare instance in which a Wald-Wolfowitz test is cited but no additional information is given, the reader has to reserve judgment as to exactly what the group differences are.

It should be noted, also, that when there are many ties in the data—that is, there are many persons with the same score—this test is inappropriate. The r value depends on a complete ranking, which cannot be uniquely determined with many persons at one score level. This test is also less powerful than several alternatives; for these reasons, it is seen in the psychological literature less often than the Mann-Whitney and Kolmogorov-Smirnov tests and may soon disappear altogether.

Table 6.14a   *Sex differences in reaction time in seconds to visual stimulus*

	Men	Women
	7	11
	6	8
	5	5
	4	2
	3	1
Median	5	5
SD	1.4	3.7

Rank position	1	2	3	4	5	6	7	8	9	10
Score	11	8	7	6	5	5	4	3	2	1
Group	W	W	M	M	M	W	M	M	W	W

A tie

$$r = 5.$$

Table 6.14b   *Age differences in reaction time (in seconds) to visual stimulus*

	10-Year-Olds	20-Year-Olds
	7	12
	6	11
	5	10
	4	9
	3	8
Median	5	10
SD	1.4	1.4

Rank	1	2	3	4	5	6	7	8	9	10
Score	12	11	10	9	8	7	6	5	4	3
Group	20	20	20	20	20	10	10	10	10	10

$$r = 2$$

6.5.2.8 The Moses test of extreme reactions. This test is used where the alternative (to H_0) hypothesis is that some experimental treatment will cause some persons to get especially high scores and others to get especially low scores. For instance, we might expect the scene of an emergency—such as a flood—to cause some persons to become panicky and others to withdraw. Anxiety scores for a group exposed to such scenes and for those not exposed are presented in Table 6.15.

The Moses test uses the span of the control-group scores as a test statistic. If the anticipated "extreme reaction" occurs in the experimental group, then, when an ordered series of all scores is made, (*a*) the control scores will tend to cluster in the middle ranges, and (*b*) the experimental scores will tend to cluster in two locations: "high" and "low." On the other hand, if the null hypothesis is correct, the various "high," "low," "middle" areas in the series will all tend to have a mix of experimental and control scores.

The number of scores, S', between and including the highest and the lowest control scores, is sometimes used as the test statistic. The sampling distribution of S' when H_0 is true is known. Tables of this distribution are consulted, and, if S' is small enough to have a probability less than the desired level of significance, then H_0 is rejected in favor of the "extreme reactions" alternative. The statistic S' is equivalent to the range of ranks in the control group. The value of S' for Table 6.15 is not significant.

Since the range is notoriously unstable, researchers may modify S' by subtracting some fixed number, h, of cases from either end of the control span; the modified span, S'_h, is then used for the test. The reader will need more detailed information to interpret a situation of extreme reactions. Most useful is a table such as Table 6.15. When the total N is quite large, a summary in the form of a frequency graph would be most helpful.

6.5.2.9 The sign test. This test is appropriate for matched-pairs designs in which there are two discrete or continuous scores for each subject or one score for each member of each of N matched pairs

Table 6.15 *Anxiety self-ratings for two movies*

Subject	Control movie	Stress movie	Sign
1	6	10	+
2	8	9	+
3	5	6	+
4	7	8	+
5	6	5	−
6	5	5	0
7	7	8	+
8	4	5	+
9	4	9	+
10	5	7	+

n.s. for Moses test; $p < .01$ for the sign test

of subjects. Each pair of scores is assigned a "$+$," "$-$," or "0," depending on whether the X_2 score is greater than, less than, or the same as the X_1 score. The number of "$+$" pairs would be about the same as the number of "$-$" scores in case the null hypothesis, H_0, of "no difference" were true.

The one-tailed test of H_0 is carried out by calculating from the binomial probability distribution with $p = \frac{1}{2}$ the probability of *as many* "$+$'s" (or "$-$'s") as the observed number. The n for this binomial distribution is the total number of plus and minus signs. In other words, 0 signs are deleted from the calculations.[4] When this probability is smaller than the desired level of significance, the hypothesis H_0 is rejected. For a two-tailed test, the probability is calculated of a number of $+$'s as deviant in either direction from the expected number, $N/2$, as the observed number.

An example of data appropriate for the sign test is given in Table 6.15. The sign test is appropriate for measures like these when the question is "Do the two conditions differ?" The two scores for each person can be readily ordered, but, due to the many scores at each score value, the entire series cannot readily be ranked without numerous ties. By converting the two scores for each person to a plus or minus, much numerical detail is lost. In this respect, it is important to note that the use of a sign test assumes that the numerical detail does not contain accurate information about the trait being assumed. The next test is definitely preferred when the magnitudes of the differences can be ranked.

6.5.2.10 Wilcoxon matched-pairs signed ranks test. This test is used for matched pairs for whom the differences between the two scores can be ranked across cases. Ranks are assigned according to the absolute values of the differences and then the sign of each difference is assigned to the corresponding rank. Zero differences are either dropped from the analysis or treated as suggested in Footnote 4, below. An example of appropriate data are given in Table 6.16. If there is no difference between the first and second scores, then the sum of the negative-sign ranks will be about the same as the sum of the positive-sign ranks. Both of these sums are determined and the test is carried out by reference to the tabled sampling distribution of the smaller.

6.5.3 Tests for Three or More Groups

6.5.3.1 The chi-square test for three or more independent groups. This test is a simple extension of the chi-square test for two inde-

---

[4] Many statisticians object to this procedure on the grounds that differences of zero actually support H_0. They recommend that zeros be assigned to the "$+$" and "$-$" categories in a manner that makes rejection of H_0 more difficult.

Table 6.16 *Observer ratings of subject's anxiety under two conditions*

Subject	X_1 (prior to stress interview)	X_2 (during 5-minute stress interview)	$X_2 - X_1$ (difference)	Signed rank
1	5	6	+1	+1.5
2	7	7	0	Delete
3	4	3	−1	−1.5
4	3	9	+6	+6
5	8	6	−2	−3
6	2	6	+4	+5
7	4	7	+3	+4

pendent groups. A table is set up in which rows (or columns) represent the groups; columns (or rows) represent the categories of the other variable. It is like a contingency table, with one important difference: the marginal frequencies for the groups are fixed, although the other dimension's marginal frequencies are free to vary. (In a true contingency table, both sets of marginal frequencies may vary.) The null hypothesis asserts that there is no relation between group membership and status on the other variable. The alternative hypothesis is that such a relationship does exist. An example is given in Table 6.17, where rows represent treatments and columns refer to certain qualitatively distinct behaviors.

The chi-square statistic is calculated from Equation 6-3. As for any contingency table, the degrees of freedom are equal to $(r - 1) \times$

Table 6.17a *Reactions to the experimental interruption of a block-building task*

Group	Reaction to task disruption			Total (N = 100)
	Start over	Abandon task	Modify task	
I: Achievement oriented instructions	15	5	5	25
II: Money incentive instructions	7	3	15	25
III: Social reinforcement oriented instructions	12	1	12	25
IV: Neutral instructions (control)	5	10	10	25
TOTALS	39	19	42	100

$\chi^2 = 19.65$ $df = 6$ $p < .01$ Contingency coefficient $= .40$

Table 6.17b $O - E$ *deviations from Table 6.17a*

Group	Start over		Abandon task		Modify task	
I: Achievement	4.75		.25		−5.50	
		(1)		(2)		(3)
II: Money incentive	−2.75		−1.75		4.50	
		(4)		(5)		(6)
III: Social reinforcement	2.25		−3.75		−2.50	
		(7)		(8)		(9)
IV: Neutral instruction	4.75		5.25		−.50	
		(10)		(11)		(12)

$(c - 1)$ where r is the number of rows and c the number of columns. A significant chi-square value indicates that group membership and the other variable (it is response to disruption in Table 6.17) are related.

A more detailed statement would have to be based on a close examination of the frequencies in the table. Table 6.17b has been prepared from Table 6.17a to aid this process. It gives the difference, $O - E$, for each cell. We see that the largest of these deviations occur in five cells, numbers 1, 3, 6, 10, and 11. Many more than expected of the achievement-oriented instruction group chose to start the task over, and many fewer than expected of that group modified it. It would seem that aroused achievement motive (if that is what existed) leads people to maintain high standards and not to compromise their efforts by task modification. This argument could be extended to cell 6, which can be interpreted as showing that money incentive induces people to get the task done, but not necessarily in the prescribed way. The moderate deviation in the cells for the social-reinforcement group lend themselves to interpretation in terms of a similarity of achievement and social motivation. Obviously, a completely detailed discussion of the tabled results would require a more thorough understanding of the experimental instructions, the study's theoretical rationale, the task used, and the persons who served as subjects. This information is usually available in written reports and should be used by the reader (and, of course, the article's writer) in formulating convincing conclusions.

6.5.3.2 The extended Median test. This test for three or more independent groups is exactly analogous to the median test for two groups. That is, a frequency distribution for all of the N scores is prepared and the median determined. Then a $2 \times k$ table is prepared with rows representing "above combined median" and "at or below combined median" and columns representing independent groups. A χ^2 statistic with $k - 1$ degrees of freedom is calculated for this table and referred to a table of the appropriate chi-square distribution. The null hypothesis is that the k groups do not differ with respect to the measured attribute. A rejection of H_0 implies the *two-tailed* alternative that they differ.

6.5.3.3 Kruskal-Wallis one-way analysis of variance by ranks. This test is used when ordinal rankings, discrete scores, or continuous scores are available for three or more independent groups and the null hypothesis states that they do not differ in central tendency. An example of such a situation is given in Table 6.18. All N of the scores are ranked in a single series. The sum of the ranks is found for each

Table 6.18a *Number of nominations as "best friend" or "friend" for three groups of girls in a sixth-grade class divided according to reading ability*

Poor ($n_1 = 4$)	Average ($n_2 = 7$)	Superior ($n_3 = 3$)
1	5	1
0	4	0
2	3	3
1	6	
	5	
	7	
	8	
4	7	3

Table 6.18b *A ranking of all scores in Table 6.18a*

Rank	Tied rank values	Score	Rank	Tied rank values	Score
1 ⎫	1.5	0	10 ⎫	10.5	5
2 ⎭		0	11 ⎭		5
3 ⎫		1			
4 ⎬	4.0	1	12		6
5 ⎭		1	13		7
6		2	14		8
7 ⎫		3			
8 ⎭	7.5	3			
9		4			

Table 6.18c   *Translation of scores in Table 6.18a to ranks*

	Poor readers	Average readers	Superior readers
	4	10.5	4
	1.5	9	1.5
	6	7.5	7.5
	1.5	12	
		10.5	
		13	
		14	
Sum of ranks	13	60.5	13
Average ranks	3.25	8.65	4.33

of the groups and a number, H, proportional to the variance of these sums, becomes the test statistic. The null hypothesis for this test is that the sums of ranks for the several groups are the same. Thus, if the groups do not differ, they will each have about the same proportion of high, low, and intermediate ranks as the others; in such a case the sums-of-ranks will have low variance. On the other hand, when the average rank for one group is higher than that for either of the others (as in Table 6.18c), the variance of the averages will be great and the null hypothesis rejected as it is here.

For k groups, the chi-square distribution with $k - 1$ degrees of freedom is the sampling distribution of H and is consulted to determine significance. A procedure has been developed by Dunn-Rankin and King for performing multiple comparisons following the Kruskal-Wallis test.[5] These follow-up tests may be one-tailed if prior predictions as to direction have been made. Tables 6.18a, 6.18b, and 6.18c give the original data, the ranking, and resultant rank data for a hypothetical application of the Kruskal-Wallis procedure.

6.5.3.4 The Cochran Q-test. This test is useful for situations in which there are k dichotomous classifications per individual—in other words, it applies in "subjects × treatments" designs in which K treatments are all given to each subject and in which the dependent variable is scored as either a success or a failure. An example is given in Table 6.19. There, each subject is classified as having passed or failed each of five test items. The null hypothesis for the Cochran test is that the proportions of pass scores are the same for all five items. The alternative hypothesis—two tailed—is that the items differ in difficulty (proportion of passes).

[5] Dunn-Rankin, P., and King, F. J., Multiple comparisons in a simplified rank method of scaling. *Educational and Psychological Measurement*, 1969, *29* (2), pp. 315–329.

Table 6.19   *Pass-fail data on six intelligence-test items*

Subject	Item					Score
	1	*2*	*3*	*4*	*5*	
1	+	+	+	+	−	4
2	−	−	−	+	−	1
3	−	−	−	+	−	1
4	−	+	−	−	−	1
5	+	+	−	−	−	2
6	−	+	−	−	−	1
7	−	−	−	+	−	1
8	+	+	−	−	−	2
9	+	−	−	+	−	2
10	+	−	−	+	−	2
Number passing	5	5	1	4	0	
Proportion passing	0.5	0.5	0.1	0.4	0.0	

$$Q = 9.17 \qquad df = 4 \qquad p < 0.10$$

The test statistic, Q, for this test is proportional to the ratio of the variance in "proportion of passes" (across persons) for items to variance in proportion of passes (across items) for individuals. The Q statistic is distributed as chi square with $k - 1$ degrees of freedom; thus, a table of such a chi-square distribution is consulted and p determined. When p is less than 0.05 or some other preset level of significance, H_0 is rejected. For the Table 6.19 example, H_0 is in fact rejected and the alternative hypothesis of differing difficulty accepted. It appears that items 3 and 5 are considerably more difficult—have lower proportions passing—than the others. The 15 pairwise differences among the several items may be tested for significance by a procedure developed by Marasculo and McSweeney.[6]

When a Cochran Q-test is used, the reader will usually be given only the proportions passing for each item (the bottom line of Table 6.19a), and perhaps, the magnitude of Q and p. This information is sufficient for answering questions about item differences in proportions of persons passing. However, if inter-item correlations are desired, a contingency coefficient must be employed (see Section 6.5.4.5). The complete information in the relevant two columns of Table 6.19a is used in calculating it.

[6]Marasculo, L. and McSweeney, M. E., Nonparametric post hoc comparisons for trend. *Psychological Bulletin*, 1967, *67* (6), pp. 401–412.

Table 6.20   *Data for a Friedman two-way analysis of variance by ranks. Entries are rankings of speakers by subjects*

Subject	Speaker 1	Speaker 2	Speaker 3
Joe	1	2	3
Bill	2	1	3
Mary	1	2	3
Jane	1	2	3
Judy	1	3	2
Alan	2	1	3
Sum of ranks	8	11	17

Test statistic $= \chi_r^2 = 5.00$ $p < .10$ $df = 2$

6.5.3.5 Friedman two-way analysis of variance. This test is appropriate when the data are (*a*) ranks or (*b*) scores for which ranks have been assigned and when there are k such ranks under k different conditions for each of N subjects. In other words, it is for the "subjects $\times$ treatments" design and ranks as data. The k scores for each subject are ranked from 1 to k; these ranks are recorded in a table such as Table 6.20 for which $k = 4$ and $N = 7$. Under the null hypothesis, there is no difference among the k conditions; thus, under H_0, the sums of ranks within conditions across subjects (column sums in the table) would be about the same.

The test statistic used is proportional to the variance of the sums of ranks. For sufficiently large N, it follows the chi-square distribution with $k - 1$ degrees of freedom; this distribution is consulted for determination of significance. A large variance among sums of ranks leads to rejection of the null hypothesis, whereas a small variance leads to acceptance. Whenever the sums differ, the variance is large; whenever they do not differ much, the variance is small. Thus, as in any analysis of variance, the hypothesis about differences is translated into one about variance. When the null hypothesis is rejected, the Marasculo-and-McSweeny[7] procedure for multiple comparisons may also be employed following the Kruskal-Wallis test.

6.5.4 Nonparametric Measures of Association

Several types of coefficients indicating the degree of association have been developed for use with scores for which the assumptions underlying the use of the Pearson correlation coefficient are not tenable. These are presented below.

[7] See Footnote 6, page 203.

6.5.4.1 Rho. Spearman's rho coefficient is equivalent to a Pearson product-moment correlation computed from ranks. The resultant rho is an index of the degree of association (from 0 to 1 in absolute value) between two sets of rankings. However, if the original scores are not ranks, rho based on a ranking of these scores will usually not be the same as a Pearson r computed from the same data, since the Pearson r gives weight to the absolute magnitude of the scores, whereas the Spearman rho considers only the ordinal position of each score. This statement is true despite the fact that the Pearson formula can be used to find the value of the Spearman rho by entering rank values into it. The differences between the r and the rho are illustrated in Table 6.21. Notice that Person 1's two scores contribute to positive correlation for the calculation of r but do not so contribute to rho. That is, both scores are rank 10; thus the difference in ranks is 0, but the numerical difference is large. The question of which coefficient is correct cannot be judged by a reader who may not know how the scores were obtained. The reader can keep in mind that rho gives degree of association for rankings and r gives degree of association for numerical scores. It is possible, thus, to have a high rho showing good correspondence between persons' rank positions on X and their rank positions on Y, but a moderate r showing a less strong ability to predict the numerical value of someone's Y score from his X score. This would happen if the relationship were monotonic and/or curvilinear. We should also note that many ties on either X or Y tend to put an upper limit less than 1.00 on rho. Thus the coefficient is not useful with ratings or other ordinal measures containing many

Table 6.21 *Correlations between time-to-finish measures on two tasks*

Person	Time score X (seconds)	Time score Y (seconds)	Rank X	Rank Y
1	21	11	10	10
2	20	7	9	8
3	16	9	8	9
4	15	5	7	6
5	14	6	6	7
6	13	4	5	4
7	12	3	4	3
8	4	2	3	2
9	3	5	2	5
10	2	1	1	1
	$r = .71$	$\rho = .90$	$\tau = .78$	

ties. The sampling distribution of rho is known, and thus hypothesis about the value of rho can be tested.

6.5.4.2 The tau coefficient. The Kendall tau is also an index of the degree of association between two sets of ranks, which can also be used with scores that are discrete or continuous. Tau is based on an examination of the order of X and Y scores within each pair. The total number, S^+, of pairs of persons is counted for which X and Y are ranked in the same order—i.e., both have $X > Y$ or both have $Y > X$. Then the number, S^-, for which the order is different for one member than for the other is counted. Tau is defined by the ratio

$$\tau = \frac{S^+ \times S^-}{P} \quad \text{where } P \text{ is} \quad \frac{N(N-1)}{2} \tag{6-5}$$

the total number of pairs in a sample of N persons. That is,

$$\tau = \frac{S^+ - S^-}{\frac{1}{2}N^* (N^* - 1)} \tag{6-6}$$

where N^* represents the number of persons left after those for whom $X = Y$ are eliminated from the analysis.

Tau has a known sampling distribution; thus, hypothesis tests concerning the true value may be made. There is also a procedure for computing partial tau coefficients.

One property of tau is that it is not on the same numerical scale as rho and r. This means that tau and rho for the same data will not be the same. For example, tau for the data of Table 6.21 is 0.78. What is important for the reader here is the idea that he must not draw conclusions based on direct comparison of a tau and a rho from the same or similar data. Tau is usually lower than the corresponding rho, and in this sense can be considered a different kind of estimate of the degree of association present.

6.5.4.3 The gamma coefficient. There are many research situations in which a measure of association is needed for two sets of discrete or continuous scores that contain numerous ties. (Ranks and other types of scores fit in this category.) The *gamma coefficient* is appropriate for correlating ordinal measures with many ties. It is a modification of the Kendall tau, in which all individuals for whom $X = Y$ are eliminated from both numerator and denominator. Thus

$$\gamma = \frac{S^+ - S^-}{S^+ + S^-} \tag{6-7}$$

A common situation wherein gamma is useful is illustrated in Table

Table 6.22 *Data for inter-rater agreement check illustration of gamma coefficient*

		Rater X's ratings of aggression				
		1	2	3	4	5
	5			2	3	2
Rater Y's ratings of aggression	4			1	5	3
	3			4	2	1
	2		2		1	
	1	1		1		

$N = 30$ $\gamma = 0.61$

6.22: X and Y represent the ordinal ratings of two observers who watched the same set of N children. Each subject is assigned to a "cell" on the basis of X's rating of him and Y's rating of him. The resultant gamma coefficient measures to what extent the two children of each pair are ordered the same way by Y as by X. Those children for whom X and Y agree perfectly are not included in the arithmetic. Thus, we ask to what extent would the children in the sample be ranked in the same way by X's ratings as by Y's?

Other situations in which gamma is useful are: test-retest correlations of ordinal rating scales and the correlation of two distinct variables, both rated on ordinal scales. Like tau, gamma is subject to varying upper limits and cannot be interpreted on the same scale as an r from the same data.

6.5.4.4 The phi coefficient. The phi coefficient is an index for the degree of association between two dichotomous variables. It is computed from 2×2 tables by the formula

$$\phi = \frac{AC - BD}{\text{Row}_1 \, \text{Row}_2 \, \text{Col}_1 \, \text{Col}_2} \tag{6-8}$$

where A, B, C, D and rows and columns refer to the frequencies indicated in the example in Table 6.23. For any given set of four marginals there are a limited number of values which phi might possibly assume. Often the upper limit—called "phi max."—of these is considerably less than 1.00. For this reason the reader may see a corrected coefficient based on phi max. used. If so, it is computed as

$$\phi' = \phi/\phi \max$$

This adjusted phi is on a scale from -1.00 to $+1.00$. The most common application of the phi coefficient is as an entry into tables to be subjected to factor analysis. The problems of interpretation that can arise from this use are quite complex and beyond the scope of this book. Hypothesis tests of the magnitude of phi are rarely made. This is because a test of the customary null hypothesis (no relationship) can be, and often is, made by performing a Fisher exact probability (small N) or chi-square test (larger N) for two independent samples. The hypothesis that rows do not differ with respect to columns (or vice versa) is just the same as the hypothesis of no association. A rejection of this null hypothesis implies that some (nonzero) association exists. (See Table 6.23. For further illustration, look back at Tables 6.10b and 6.11 and their associated values of phi.)

6.5.4.5 The contingency coefficient. The contingency coefficient is based on the chi-square statistic. It is useful as a descriptive measure of the degree of association between two variables, each involving several nominal categories. From a contingency table for the two variables, the chi-square statistic is computed—as for an hypothesis test. Then, in order to make it vary between 0 and 1, the chi square, χ^2, is entered into the following formula for the contingency coefficient, C, where

$$C = \sqrt{\chi^2/(N + \chi^2)} \tag{6-9}$$

The contingency coefficient often has an upper limit in absolute value less than 1.00. The exact magnitude of this upper limit depends on the numbers of rows and columns.

This chapter contains two contingency tables (Tables 6.8a and

Table 6.23 *Data for a phi coefficient*

		Test item 2		
		Pass	*Fail*	
	Pass	10	5	15
		A	B	
Test item 1				
	Fail	6 C	9 D	15
		16	14	
$\phi = .267$		ϕ max. $= .285$		

6.17a), which were introduced in connection with discussion of chi-square tests. Each of them has a moderate contingency coefficient. Note that when the null hypothesis (of no association) is rejected, the contingency coefficient describing the degree of association may still be modest in magnitude. A statistically significant relationship may turn out not to be significant in any useful way. Whether or not such a coefficient is useful depends on considerations outside of just the one calculation. The coefficient should be examined in relation to other coefficients calculated from the same subjects for other, theoretically relevant, variables.[8]

6.5.4.6 The Kendall coefficient of concordance. The Kendall coefficient of concordance is useful when three or more sets of rankings or ranks substituted for discrete or continuous scores are available for N subjects. The question is, "Is there some consistency of ranking across treatments (or measures)?" Suppose, for example, that seven subjects are given four mazes; the subjects are ranked according to the order of finishing for each maze. The results might appear as shown in Table 6.24, or they might just be given in a written summary. Each person has three ranks representing his rank on Maze 1, Maze 2, and Maze 3. The coefficient of concordance, W, is directly (linearly) related to the variance of the k sums of ranks, where k is the number of conditions.

Tests of hypotheses concerning W may be made, since (*a*) exact probabilities can be worked out for small N's, and (*b*) W follows a chi-square distribution for sufficiently large N's. An especially useful application of W is to situations in which one wishes to assess agreement among several observers; in such an instance, the k observers constitute the k treatments.

[8] For an extended discussion of the contingency coefficient, the reader is referred to McNemar, Q., *Psychological Statistics*, New York, Wiley, 1965, pp. 227-231.

Table 6.24 *Order of finishing for four maze tasks*

Subject	*Maze 1*	*Maze 2*	*Maze 3*	*Maze 4*	*Sum of ranks*
Joe	1	2	3	1	7
Jim	3	4	4	2	13
Bill	2	1	1	5	9
Ron	5	3	2	4	14
Phil	7	7	7	6	27
Frank	4	5	6	3	18
Paul	6	6	5	7	24

Coefficient of concordance, W, $= .74$

7

The Reader's Task

In this book, techniques and guidelines have been offered for the critical reading of the research literature. This concluding chapter offers both a checklist summary of evaluations to be made in perusing an individual article and a set of suggestions for comparing several articles.

7.1 A Checklist for the Individual Article

The following checklist assumes you are examining an article because it seems relevant to a topic you are interested in. The writer of a particular article may have planned and conducted the study in order to demonstrate the applicability of the verbal operant-conditioning paradigm to aggression. A professor of developmental psychology who is preparing a series of lectures on children's aggression might read it because she wants to cite an experimental investigation of verbal operant conditioning of aggressive responses. A student might read the same article because he is looking for sources of information for a term paper on behavior-modification techniques.

Whatever your purpose might be, you will have to evaluate the article with regard to both the author's purpose in conducting the investigation and your own purpose. Since no author should be held to account for not considering a reader's goals, the adequacy of any design and set of measures should be judged in terms of the author's purposes. The decision whether or not the article will be useful for you is not so much a judgment of the quality of the author's research as it is a judgment of relevance to your needs. This book has been addressed to critical readers who want to make both kinds of judgments—judgments of adequacy in terms of the investigator's own purposes *and* judgments of relevance to their own needs.

7.1.1 The Explicit and Implicit Hypotheses

The author of an article will usually identify the hypothesis (or hypotheses) of interest to him and state it (or them) explicitly. For example, the study on verbal operant conditioning of aggression might employ two main groups of children: (*a*) an experimental group receiving social reinforcement for selecting an aggressive verb in a five-alternative multiple-choice situation and (*b*) a control group reinforced at random. If this were so, the author would state as an alternative to the null hypothesis his research hypothesis: the experimental group will have a higher mean for "number of aggressive responses" than will the control group.

You may well find this explicit hypothesis of interest. But, in addition, you may be interested in hypotheses relating to differences of sex, age, or social class. Perhaps the specific nature of the four nonaggressive alternative responses are of interest; tests of these additional hypotheses may be included in the investigation. If the additional variables—sex, age, social class, etc.—were controlled. For example, if the main experimental groups were subdivided for two ages, two classes, and the two sexes, and a factorial analysis of variance were done, tests may well have been made of these other hypotheses. In other experiments, there will be no tests of the other hypotheses, in which case they can be examined only partially and indirectly. No hypothesis that is not specifically tested can be rejected with any known degree of confidence.

You should compare the explicit hypothesis or (hypotheses) of the investigator with the other hypotheses of interest to you. How much similarity is there between behavior modification of aggression and the study of verbal operant conditioning of aggression? The investigator's research hypothesis—that "social reinforcement of aggression increases the frequency of aggression"—is quite similar to his more general hypothesis—that "aggression is behavior subject to modification." If you are interested in a hypothesis much different from that of the investigator, the usefulness of the article to you will depend on whether or not the sample is divided according to variables of interest to you either (*a*) because they are control variables or (*b*) because they provide descriptive categories.

7.1.2 Measures of the Variables

A research article will usually describe the procedures used to measure each subject's status on all the variables under investigation. In an experimental investigation, these will be categorized as independent and dependent variables. You should carefully note these definitions as well as the level of measurement of each. Conclusions made either by the author or by you regarding any of these variables should not extend beyond the scope of the definitions. For example, the hypothetical study about the verbal operant conditioning of aggression should stimulate conclusions only about verbal multiple-choice responses. Generalizations about the determinants of aggression in general or of verbal aggression in general could not be based on this one study alone.

7.1.3 Population Sampled

The relevance of any given study to your purposes depends in part on the extent to which the sample of subjects can be considered representative of the population about which generalizations are to be made. You need first to characterize the population that interests you in terms of the important demographic characteristics of its members, such as age, social class, school grade, occupation, and so forth, and also in terms of personality traits that should be sampled. For example, our developmental-psychology professor may have an interest in the population of five-year-old middle-class girls of normal intelligence. Or she might be interested in several populations defined in terms of age and sex, such as middle-class five-year-old boys, middle-class five-year-old girls, middle-class ten-year-old boys, middle-class ten-year-old girls, and so forth.

The next step, after identifying the population of interest, is to identify the population actually studied by the investigator. In some instances, this population will be clearly labeled as "all the children enrolled in *X* unified school district." In other instances, the population will have to be inferred from the description of the sample, as, for example, when the author states that thirty four-year-olds from "local preschools" served as subjects. Either way, it is rare for all of the prominent characteristics of the population to be identified. Specifying some of the characteristics may require some familiarity with the locale in which the study was done. Your conclusions about what population has been sampled in a given study must be considered tentative. (This problem was discussed in Chapter 3.)

After both the population of interest and the population actually studied are identified, they should be compared. The degree of relevance of the article to your purposes will depend in part on the degree of similarity between these populations. To the extent that these populations differ with respect to important variables such as age or sex (percentages of males and females) or social class, or IQ, the applicability of the study's findings to your problem is limited. Most careful readers would not wish to make generalizations about causes of behavior based on results from studies done only with adult female mentally retarded persons. These same readers might be more willing to generalize from university laboratory preschoolers to all bright four-year-old children. Resolving questions about the identity of certain populations and generalizing among them involves much careful judgment.

7.1.4 Descriptive Statistics

After considering the hypothesis, measures, and population, you should turn your attention to the summary of descriptive statistics. For each variable being studied, you should note the statistic(s) for central tendency (often mean or median), as well as the extent of variability. These statistics give you an indication of what the normal or usual range of scores is. The statistical tests performed by the investigator make direct use of these descriptive statistics. You can use them to gain additional insight. Consider as an example the verbal operant-conditioning experiment mentioned earlier.

The score might be the number of aggressive choices out of fifteen opportunities. The descriptive statistics might be these—

Statistic	Control	Experimental
Mean	7.2	10.1
Standard deviation	2.8	1.9
N	15	20

A two-tailed *t*-test of the hypothesis of no control versus experimental mean difference would lead to its rejection. You can further note that the variability for the control subjects was greater than that for the experimental subjects—the implication being that there is some, but not considerable, overlap between the distributions. The mean of the control distribution falls, for instance, at the point on the experimental distribution that would be the seventh percentile of the experimental distribution if it were normal. The examination of the descriptive statistics confirms that the statistically significant difference is also substantial.

Here are the descriptive statistics for the hypothetical replication of the study on which the previous table was based; in this case, there are 100 cases per cell.

Statistic	*Control*	*Experimental*
Mean	7.5	7.9
Standard deviation	2.5	2.5
N	100	100

These figures indicate much overlap of the experimental and control distributions and yet the mean difference of .4 of a point attains significance. In this instance, you might justifiably decide that the difference was too small to warrant use in developing theory about verbal conditioning. There is a range of opinion among professional behavioral scientists and others concerning the use of small but significant differences in a theoretical context. The important point, in the context of the present book, is that descriptive statistics do provide information in addition to that contained in the outcome of statistical tests.

7.1.5 *The Design of the Study and Statistical Analysis*

After acquainting yourself with the hypotheses—as well as the population sampled and the measures used—you will want to evaluate the design of the study in relation to the hypotheses being tested. The groups employed and the associated definition of variables in terms of group membership should be examined. Do the groups adequately define the variable(s) under study?

How is the hypothesis to be tested? Are two or more groups to be compared or is some sort of correlation coefficient to be calculated. Are there adequate controls for other variables—such as traits of the subject—that might influence the outcome?

7.1.6 *The Conclusion and Discussion Section*

Finally, you will want to compare the statements made by the author about his data with the data and the statistical analysis of it. Detailed suggestions for how this comparison can be done have constituted the main subject matter of this book.

7.2 Comparing and Relating Two or More Research Articles

Readers who are preparing review-of-the-literature papers or who are attempting to master a substantive topic in psychology may be faced with the task of comparing studies that have been done for diverse purposes. Similarities and differences between articles can arise under each of the heads on the checklist for individual articles. That is, two studies may have (1) similar or different hypotheses, (2) similar or different ways of measuring variables, (3) similar or different populations for sampling subjects. They may produce (4) descriptive statistics that show similar or different information; (5) they may employ similar or different research design and (6) similar or different statistical analyses.

You should proceed to consider each article with some very general hypothesis of interest in mind. The frustration-aggression hypothesis is an example. In this hypothesis "frustration" is the independent variable and "aggression" is the dependent variable. The hypothesis states that there is a high relationship between the occurrence of the frustration of an individual and subsequent aggression by him. If you are interested in this hypothesis, you will have assembled a reference list of studies dealing with frustration and/or aggression. Perhaps you will have limited your search to some part of all the studies available. Perhaps you are looking only at experimental studies, or studies done with human adults, or studies in which the subjects are preschool children. In any case, you have a group of studies to read, analyze, and integrate.

A suggested preliminary step is to read each study once in order to gain an idea of how many and what kind of different measures and different subject populations are to be encountered. Next, classify the studies according to the attributes itemized in the foregoing checklist for individual articles. Consider the general question or hypothesis (frustration-aggression). Some of the articles propose to test the hypothesis or address the question directly, whereas others provide relevant information indirectly. This is the first classification. Next, consider the different ways in which the variable frustration is operationally defined and measured—paper and pencil test, experimental inducement by means of instructions, experimental inducement by means of experimental arrangements. Reclassify the articles on this basis. Next, consider the different ways in which aggression is defined and measured and reclassify the articles.

After the articles have been classified according to their uses of measurement and the hypotheses tested, they can be classified according to the population sampled—the different representations of age, sex, social class, race, and so forth. The various results and conclusions are the basis for a final classification.

The purpose of this classification is to provide you with a basis for deciding which investigations provide findings in support of the general hypothesis (or answers to some question) and which do not. For example, is it only experimentally induced frustration which produces aggression? Is the hypothesis only true for adult males? Does favorable evidence arise only if the subjects are put in some situation that suggests aggression, such as a competitive contest?

Finally, in integrating the results of a group of investigations on the same subject, you should be sure that your own conclusions are warranted by the investigations. If the limitations of your general statements can be spelled out in terms of the characteristics on the checklist, your understanding of the topic should be enhanced. Ideas for further research projects needed in a field can also be formulated more precisely if the foregoing checklist is followed.

Index